Mechanical behaviour of ceramics

R. W. DAVIDGE
Head of Ceramic Technology Group, AERE, Harwell

Mechanical behaviour of ceramics

CAMBRIDGE UNIVERSITY PRESS
Cambridge
London New York New Rochelle
Melbourne Sydney

Published by the Press Syndicate of the University of Cambridge
The Pitt Building, Trumpington Street, Cambridge CB2 1RP
32 East 57th Street, New York, NY 10022, USA
296 Beaconsfield Parade, Middle Park, Melbourne 3206, Australia

First published 1979
First paperback edition 1980

Printed in Great Britain at the
Alden Press, Oxford

Library of Congress Cataloguing Publication Data

Davidge, R. W., 1936—
Mechanical behaviour of ceramics.

(Cambridge solid state science series)
Includes index.
1. Ceramic materials–Testing. I. Title.
TA430.D37 620.1′4 77-90206
ISBN 0 521 21915 9 hard covers
ISBN 0 521 29309 × paperback

Contents

Preface

The aims of this book are twofold: first, to present the scientific foundations underlying the mechanical behaviour of engineering ceramics; secondly to show how this theory, together with less rigorous empirical approaches where necessary, can be applied to the engineering use of ceramics. The appeal is thus both to the materials scientist and to the technologist dealing with engineering applications. The links between the materials science and engineering aspects have been insufficiently emphasised in the past. In consequence, the ceramist often regards the engineer as someone who does not know how to handle ceramics, whereas the engineer regards ceramics as fragile materials of variable and unpredictable properties. There is some truth in both of these viewpoints and it is essential that a new type of materials technologist emerges who can both understand ceramic materials and design engineering components from them.

The distinguishing mechanical property of ceramics (as witnessed by any ham-fisted dish-washer) is brittleness – that is, catastrophic failure following an almost entirely elastic deformation. Were it not for this limitation the use of ceramics for general engineering applications would be widespread; other, attractive properties such as hardness, stiffness and refractoriness could be exploited to the full. Brittleness is thus a dominant theme throughout this volume.

The crucial theoretical background stems directly from the basic theory of brittle fracture developed by A. A. Griffith in the early 1920s. It was only some 30–40 years later that these ideas were applied systematically to ceramics – first, in the form of single crystals that exhibited cleavage fracture and subsequently, to more useful ceramics. Over the last decade, however, progress has been spectacular and the science of the mechanical behaviour of ceramics, particularly strength, has now reached a consolidated stage where it is being applied ever increasingly to important engineering applications. Much of this information is still scattered in the literature and it is thus timely that the principles are gathered together in a single volume.

The first part of the subject matter (chapters 1–6) is concerned with the relationship between the mechanical properties of ceramics and their atomic bonding and microstructure. To emphasise the complexity of practical ceramics we summarise first (chapter 1) the definition and use of ceramics, their bonding and atomic arrangements and their methods of manufacture and resulting microstructures. The general stress/strain behaviour of ceramics is typified by two extremes: brittle failure at low temperatures and ductile failure at high temperatures. The elastic properties of ceramics (chapter 2) are amongst the most fundamental and important and are influenced by the type of bonding and the microstructure of the material. Elastic deformation is, however, limited, generally to ~0.1%, being followed by either fracture or flow. Fracture in ceramics (chapter 3) is controlled by the stress to propagate small, sharp flaws and can be understood quantitatively by the application of linear elastic fracture mechanics. A consideration of plastic flow in ceramics (chapters 4 and 5) leads to the conclusion that significant plasticity is generally not possible at temperatures useful for engineering purposes and that the limited flow processes available lead to the generation of new flaws. These ideas are then drawn together (chapter 6) through examination of the fracture strength of selected ceramics.

The impact resistance and the development of tough ceramics is discussed in chapter 7 and demonstrates how useful improvements to toughness can be obtained, particularly by the incorporation of fibres into ceramic matrices. At present, however, the key design parameter for engineering ceramics is the useful working strength. Thermal-stress resistance is an important topic for the high-temperature use of ceramics and the background is presented in chapter 8.

Finally (chapter 9), the development of engineering design parameters for ceramics is described, including topics of special significance to engineering components: statistical variations in strength, time dependence of strength and the effects of multiaxial stress.

It is hoped that this brief exposure to the mechanical behaviour of ceramics, apart from providing a useful summary of the state-of-the-art, will encourage both the materials scientist to develop better theories and materials, and the engineer to design ceramic components with confidence.

I am indebted to my research collaborators, without whom many of the experimental and theoretical developments described here would not have been possible. Thanks are due to R. W. Cahn for his persistence in persuading me to accept this undertaking and his continued interest and encouragement. I am grateful to a number of colleagues including J. R. McLaren, D. C. Phillips, D. Pooley, P. L. Pratt and G. Tappin, who through discussions and criticism of the text have made valuable contributions.

July 1977 R. W. Davidge

1 Background to ceramics

The properties of ceramics, as of other materials, are highly dependent on their texture, both on the atomic and on the micro- and macrostructural scales. Texture thus represents a key and central feature in the science of ceramics, linking fabrication and properties (fig. 1.1). The raw materials, the way they are processed and fired, can all affect the texture of the material and hence its properties. In addition, a number of external factors, particularly the environment, the temperature and the test conditions, are important in determining properties.

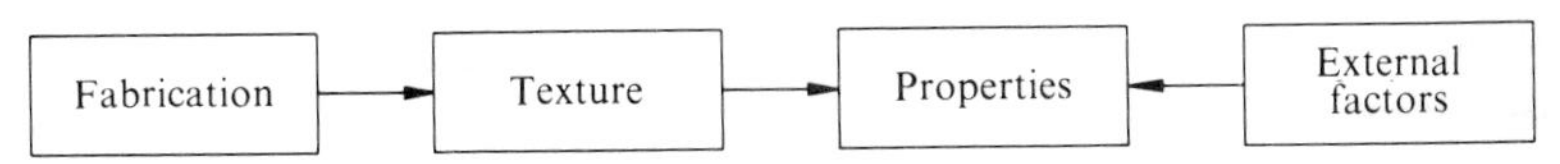

Fig. 1.1. Factors controlling the properties of ceramics.

Ideally, the ceramist attempts not only to explain the properties of interest in terms of the texture of the material but also to adjust the fabrication conditions so as to optimise the important properties. However, consideration must be given also to the form and design of the ceramic component. For traditional ceramics, such as building materials, refractories and pottery, the component is of simple shape and the engineering design aspect is not often of major importance. But for the newer class of engineering ceramics, for example high-strength oxides, carbides and nitrides, and ceramic-based composites, design is of vital importance if the properties of ceramics are to be utilised to the full; furthermore, data on mechanical properties must be expressed in a form relevant to engineering applications.

Although our primary aim is to develop an understanding of mechanical properties under a variety of physical conditions in terms of texture, it is essential to consider how ceramics are fabricated and hence the likely limits to their performance. In this chapter we outline a definition of ceramics, mention the types of atomic bonding and the geometrical arrangements of atoms, indicate briefly the ways in which ceramics are fabricated and their resulting microstructures, and introduce the responses of ceramics to stress.

1.1 Definition

The main classes of ceramic material are listed in table 1.1. Included are

Table 1.1. *Definition of ceramics*

Class	Material type	Typical uses	Importance of mechanical properties
Pottery	China, porcelain, earthen-ware, fireclay	Tableware, wall tiles, sanitary ware, electrical insulators	*
Heavy clay	Bulk clayware	Bricks, roof tiles, floor tiles, pipes	**
Abrasives	Alumina, silicon carbide, diamond	Grinding wheels, abrasive pads, polishing powders	***
Refractory	Alumina, silica, alumino-silicate, magnesite, carbon and graphite, zirconia	Furnace linings, molten-metal moulds	**
Glass	Glass, glass ceramics, enamels	Containers, flat glass, glassware, enamelled ware	*
Cement	Portland and alumina cement	Structural cement and concrete	**
Engineering ceramics	Oxides, carbides, nitrides, cermets, ceramic composites	Bearings, seals, dies, engine components	***
Electrical/optical/magnetic	Various	Capacitors, solid-state electrolytes, special windows, magnets	*
Nuclear	Oxides and carbides of fissile metals	Nuclear fuels	**

some common materials of each class and an indication of typical uses. The importance of the mechanical, as opposed to other, properties is denoted by a one- to three-star rating. The mechanical properties of all materials are of some significance in that all components are subjected to certain handling; but this is of minimum importance for, say, a ceramic magnet. At the other extreme the mechanical properties are of major significance for abrasives and engineering ceramics. There is no completely satisfactory and universally accepted definition of ceramics and the table should be regarded only as indicative of the main types and uses. More detailed information is given in Kingery *et al.* (1975) and Norton (1974).

1.2 Bonding and atomic arrangements

Ceramics are typical solids in which the atoms or ions are arranged in regular arrays, although in glasses the regularity is only short-range. The type of bonding and the atomic arrangements affect a wide range of mechanical properties including elastic constants, hardness, and plastic properties such as slip by dislocation motion. This is a vast subject but a few relevant basic principles are worth reviewing at this stage.

Bonding

Bonding in ceramics is mainly of ionic or covalent type, and usually a hybrid of these. The tendency towards ionic bonding between atoms increases with increasing difference in the electronegativity of the atoms. Electronegativity is a qualitative property which is a measure of the

Table 1.2. *Electronegativity values for elements in the first three periods of the periodic table*

H 2.1						
Li 1.0	Be 1.5	B 2.0	C 2.5	N 3.0	O 3.5	F 4.0
Na 0.9	Mg 1.2	Al 1.5	Si 1.8	P 2.1	S 2.5	Cl 3.0

power of an atom in a molecule to attract electrons to itself (Pauling, 1948). Electronegativity values can be ascribed to each element from a consideration of the bond energies between that element and other elements, and Pauling's values for the elements in the first three periods of the periodic table are given in table 1.2. Note that elements in group I of the periodic table are at opposite ends of the electronegativity scale to elements of group VII with a systematic variation between. It is clear in a qualitative sense that an increase in the difference in the electronegativity values of the two elements in simple compounds leads to an increase in the proportion of ionic bonding. Materials comprising a single group IV element such as carbon or silicon, with zero electronegativity difference, can thus be associated with pure covalent bonding. It would be useful to be able to make quantitative statements for other compounds regarding the proportions of ionic and covalent bonding. The data of the electronegativity scale provide the basis for this but the problem of calculating absolute values is not easy and remains empirical at present. Using reasonably justified assumptions about the amount of ionic bonding in various hydrogen–halogen compounds Pauling produced the curve shown in fig. 1.2. Although the values are not absolute the curve permits an approximate and comparative estimate of the type of bond, and details for selected compounds are included in table 1.3. There is a wide range of bond type, from MgO which is mainly ionic to SiC which is predominantly covalent.

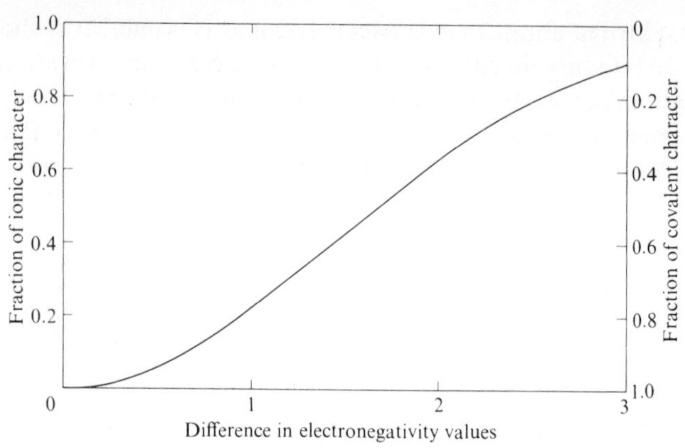

Fig. 1.2. Variation in bond character with difference in electronegativity values of constituent atoms of compounds. (After Pauling, 1948.)

Table 1.3. *Fractions of ionic and covalent bonding for simple compounds*

Compound	LiF	MgO	Al_2O_3	SiO_2	Si_3N_4	SiC	Si
Electronegativity difference	3.0	2.3	2.0	1.7	1.2	0.7	0
Fraction ionic bonding	0.89	0.73	0.63	0.51	0.30	0.11	0.00
Fraction covalent bonding	0.11	0.27	0.37	0.49	0.70	0.89	1.00

Atomic arrangements

The ways in which individual atoms in ceramic compounds are arranged geometrically depends on several factors including (*a*) the type of bonding, (*b*) the relative sizes of the atoms, and (*c*) the need to balance electrostatic charges.

In covalent crystals the nature of the bond dominates and in β-SiC each atom is surrounded by four neighbours of the other element. This structure, which is similar to that of diamond, is shown in fig. 1.3(*a*).

In ionic crystals the bond is less directional and the relative atomic sizes and charge-balance factors become of increasing importance. In most oxides the oxygen ions form close-packed arrays and the metallic ions, which are usually smaller, are arranged in the interstices. Thus, the larger the relative difference in the radius of the ions, the smaller the expected coordination number (the number of oxygen ions surrounding each metal ion). The range of radii ratios over which a particular coordination number is likely can be computed from considerations of the packing of hard spheres. For example, six-fold coordination is expected when the ratio is between 0.73 and 0.41. In MgO the radius ratio (Mg:O) is 0.47 and MgO has the cubic NaCl structure of six-fold coordination, fig. 1.3(*b*). In α-Al_2O_3 the radius ratio (Al:O) is 0.41, the value separating six-fold and four-fold coordination, and in this case a six-fold coordination is also observed but of hexagonal structure, fig. 1.3(*c*). Although these size rules are a useful guide, other factors like

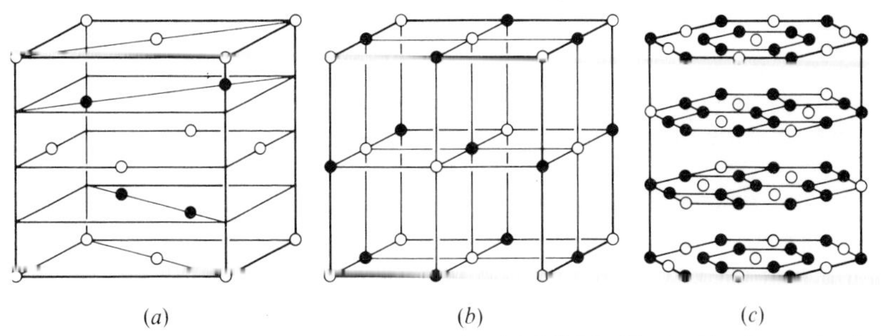

Fig. 1.3. Crystal structures of ceramics: (*a*) β-SiC (cubic zinc blende structure) ○, Si; ●, C; (*b*) MgO (cubic sodium chloride structure) ○, Mg; ●, O; (*c*) α-Al_2O_3 (hexagonal close packed structure) – only the Al atoms (●), which fill two-thirds of the available sites, are shown; ○ represents the unfilled sites; hexagonal layers of oxygen atoms are sited between the Al layers (see fig. 4.2 for further details).

distortion or polarisation of the ions and tendencies towards covalency play important roles. Further information about the diversity of possible atomic arrangements is given in Kingery *et al.* (1975) and Evans (1964).

1.3 Fabrication and microstructure

The fabrication process converts particulate raw materials into useful solid objects. This involves forming the particulate matter into the desired shape, followed by a sintering stage at high temperature. Sintering is a term used to cover the processes whereby the original particles join together, often accompanied by a reduction in volume and in the free spaces between the particles. A variety of processes is available and that chosen depends both on the intricacy of the product shape and the properties required. Our interest in fabrication is not primarily in the precise geometry of the product and how this is formed, but in the texture of the material and how this is related to the manufacturing process. It is, however, essential to appreciate the fundamentals of fabrication, with regard to sizes and shapes possible, because of the relevance to the engineering design process – the optimum design from a stress point of view may be impossible to fabricate. The most widespread fabrication process is sintering in the presence of a liquid phase but to attain the best mechanical properties it is often necessary to resort to more complicated techniques such as hot-pressing or reaction-sintering.

Fabrication processes other than sintering, which we mention just in passing, are those for glass and cement. Glass is made by melting together the constituent oxides which are then formed into products whilst the glass is in a semiliquid state. Cement is a fired mixture of clay and limestone which is then ground to fine powder; a chemical reaction occurs on mixing with water. Concrete, which is a mix of cement, sand and aggregate, is thus readily formed by pouring a fluid mix into moulds and allowing the mix to set.

Solid-state sintering

In its simplest mechanistic form, the sintering process involves heating a material formed by the compaction at ambient temperatures of fine particles (usually $< 1\ \mu m$ diameter) of a pure powder. At high temperatures, typically from 0.5 to 0.8 of the absolute melting temperature, the particles sinter together. This is a spontaneous process and must thus be accompanied by a decrease in free energy of the sample. The most important driving force for sintering is the reduction in solid/vapour surface area when the individual particles fuse together and the larger particles grow at the expense of smaller ones. Sintering is thus the reverse of fracture; the creation of new surface area by fracture requires energy as we shall see in detail later.

Idealised solid-state sintering is of restricted practical use for ceramics in that pure materials and fairly high temperatures are involved. Solid-state sintering does, however, have important applications, for example in the preparation of pure uranium dioxide for nuclear fuel, and it is possible to obtain solid material close to theoretical density. The resulting microstructure comprises individual crystallites or grains, separated by grain boundaries, and probably residual porosity. The grain size is usually much greater than the original particle size.

Hot-pressing and liquid-phase sintering

Hot-pressing refers to the simultaneous application of heat and pressure during sintering. The advantages over normal sintering are that higher densities and finer grain sizes may be achieved at lower temperatures. These result from the increased driving force for sintering caused by the stresses set up at the points of particle contact.

The main features of a hot-pressing facility include a refractory die into which is put a finely ground ceramic power and a means of applying pressure to the die at high temperature. The most common die material

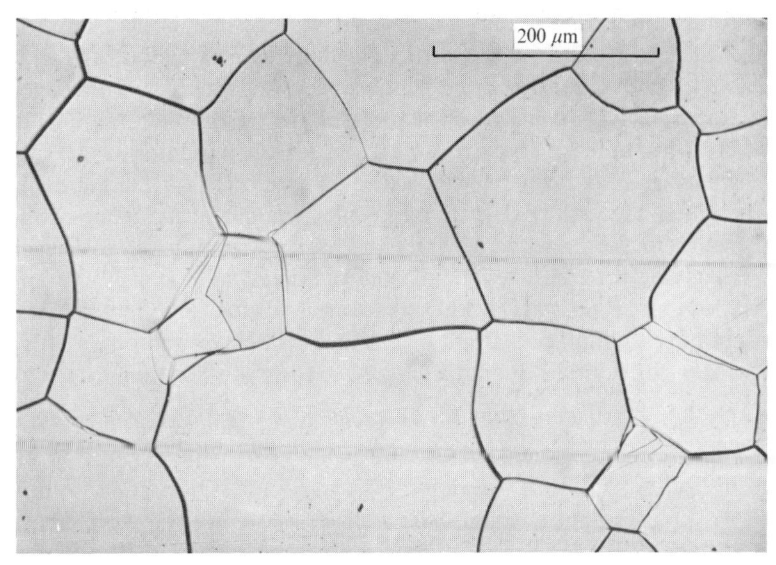

Fig. 1.4. Microstructure of fully dense, transparent magnesia. (After Davidge, 1969.)

is graphite. This needs the protection of an inert atmosphere and the strength of graphite usually limits the available pressures to ~ 50 MN m^{-2}. The times required for pressing are usually quite short (a few minutes) and most of the fabrication time is taken in heating and cooling the assembly.

Because of the expense and difficulty of working at high temperatures, it is often advantageous to include small, controlled amounts of additives that form plastic phases which enable sintering to take place at somewhat lower temperatures. For example, small amounts of magnesia greatly assist the sintering of both alumina and silicon nitride. In some cases a fugitive additive can be used, such as lithium fluoride in magnesia, which can then be removed by a subsequent heat treatment. The microstructure of this hot-pressed magnesia is shown in fig. 1.4. This is a single-phase material with a uniform grain size and virtually no residual porosity. It should be noted that the advantages conferred by controlled chemical additions during hot-pressing may have to be balanced against the disadvantage of a reduction in refractoriness.

Liquid-phase sintering in traditional ceramics

The traditional pottery and clayware ceramics are shaped by the familiar forming processes, rendered possible by the plasticity of clay. After forming, the ware is dried, and sintered to give a mixture of fused oxides.

Pottery bodies are usually a blend of three constituents: clay, flux and filler. The mix retains much of the plasticity associated with the clay. Table 1.4 gives some mix proportions for several types. Clays are mixed hydrated oxides, mainly of silica and alumina, with minor amounts of several other oxides. A typical pure clay is kaolinite $Al_2O_3 . 2SiO_2 . 2H_2O$. The particle size of clays is very small, ~1 μm across and 0.1 μm thick. The atoms are arranged as alumino-silicate layers in the plane of the platelets. In the presence of water the particles readily slip over each other giving the characteristic plasticity. Fluxes, such as the feldspars typified by $K_2O . Al_2O_3 . 3SiO_2$, are anhydrous alumino-silicates, containing sodium, potassium or calcium. Fillers are more refractory particles, for example quartz, a crystalline form of silica.

Table 1.4. *Proportion (volume %) and particle size of materials in some typical pottery bodies. (After Dinsdale and Wilkinson, 1966)*

Material		Approx. median size (μm)	Earthen-ware	Bone china	Sanitary fireclay	Sanitary whiteware	Electrical porcelain	Tile
Clay	Ball clay	0.5	25	–	–	23	25	22
	China clay	1.5	25	25	–	24	25	22
	Fireclay	15	–	–	60	–	–	–
Flux	Stone	10	15	30	–	–	–	15
	Feldspar	10	–	–	–	20	25	–
Filler	Flint	10	35	–	–	33	25	41
	Bone	5	–	45	–	–	–	–
	Grog	1000	–	–	40	–	–	–

On heating the blended materials, the water and other volatiles evaporate first. At temperatures ~1000 °C and higher the clay particles and fluxes begin to react together to form a predominantly glassy mass and at this stage sintering and densification occur. This sintering process is complete at typically ~1200 °C. The filler particles being more refractory play a relatively minor chemical role in the sintering process, although some dissolution into the glassy phase occurs. Important aspects of the filler are that it reduces the total shrinkage and imparts rigidity to the body during firing so that the original geometrical form is preserved (albeit on a smaller scale).

The microstructures of traditional ceramics, even relatively simple ones, tend to be complex. Figure 1.5 illustrates that of a simple experimental ceramic made from 50% china clay, 30% alumina (an inert filler) and 20% nepheline syenite (a flux). The white alumina particles are relatively unreacted. The background matrix is mainly glassy but does contain crystalline phases such as mullite ($3Al_2O_3 . 2SiO_2$). Also obvious are the small black regions which are residual porosity.

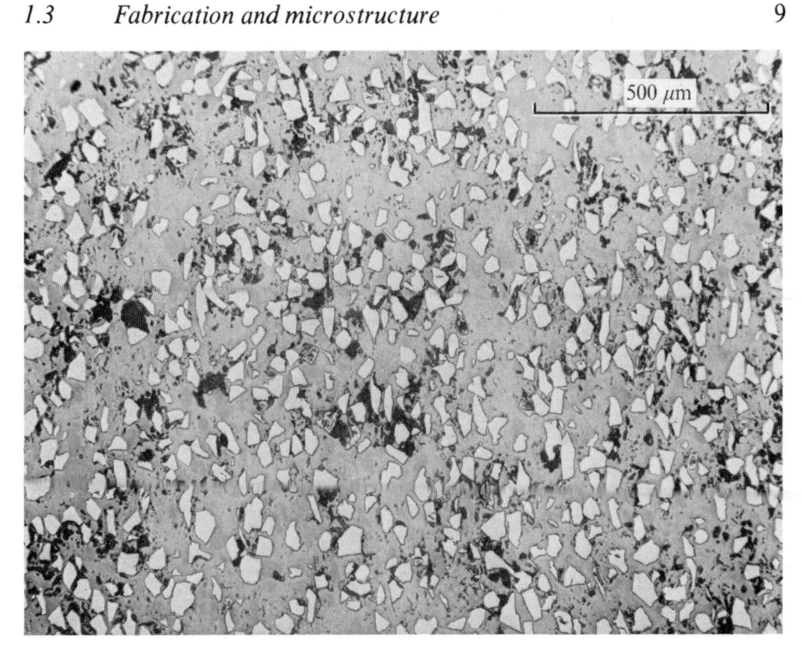

Fig. 1.5. Microstructure of experimental clay-based ceramic showing alumina particles in predominantly glassy matrix.

Although the formulations for other traditional ceramics such as refractories and vitreous-bonded abrasives are very different from those described above, the sintering process is based on the same principles. Refractories are made by firing compacted particles of the major phase(s) which usually contain small amounts of impurities which act as fluxes. Similarly for porous vitreous-bonded alumina abrasives the hard ceramic particles are pressed together with small amounts of clay and flux, which again fuse at high temperatures.

Reaction-bonding

This process, of increasing technological importance, involves sintering by way of a chemical reaction. This is not a logical route for oxide production because it would require reduction of oxide ores to metal and then oxidation back to the oxide. The technique is used particularly for silicon-based ceramics such as silicon nitride and silicon carbide. The following reactions occur at high temperatures,

$$\begin{aligned} 3Si + 2N_2 &= Si_3N_4 \\ Si + C &= SiC \end{aligned} \tag{1.1}$$

and bonding occurs simultaneously.

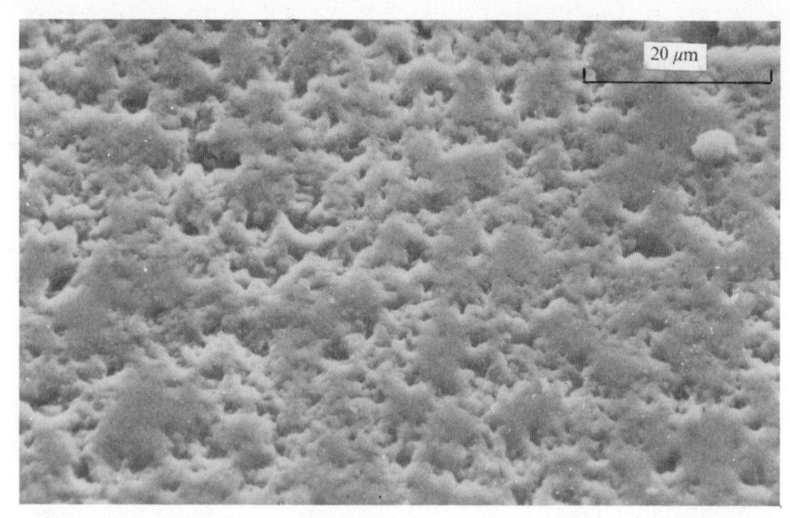

Fig. 1.6. Scanning electron micrograph of polished surface of reaction-bonded silicon nitride revealing porosity. (After Davidge *et al.*, 1972.)

To produce reaction-bonded silicon nitride, the articles are first formed from silicon powders by a variety of processes. These include flame-spraying of silicon on to a former – this process is useful for shells and hollow structures; isostatic pressing of silicon powders; or mixing of silicon powder with plastic binders and then plastic-forming using techniques such as rolling, extruding or pressing. The silicon content at these green stages is ~50%, the remainder being plastic additives or voids. After burning out the additives where appropriate, at low temperature, the components are then fired in a nitrogen atmosphere. This is generally performed partly below the melting point of silicon (1410 °C) and then, once a skeleton of silicon nitride has formed, above the silicon melting point. The total firing schedule is over a few days. The resulting microstructure, fig. 1.6, comprises a pure, fine-grained silicon nitride. The porosity is generally in the range 10–30%. Material containing < 10% porosity cannot readily be produced in convenient firing times because the porosity is necessary to allow access of the nitrogen to the centre of the component.

Silicon carbide could be made, in principle, using a similar technique by reacting particles of carbon with liquid or gaseous silicon. However, it has been found advantageous to start with a mixture of silicon carbide particles plus a fine graphite powder before reacting with silicon. The silicon carbide and graphite are blended with plastic binders and formed into shapes in similar ways to those described for silicon nitride. After

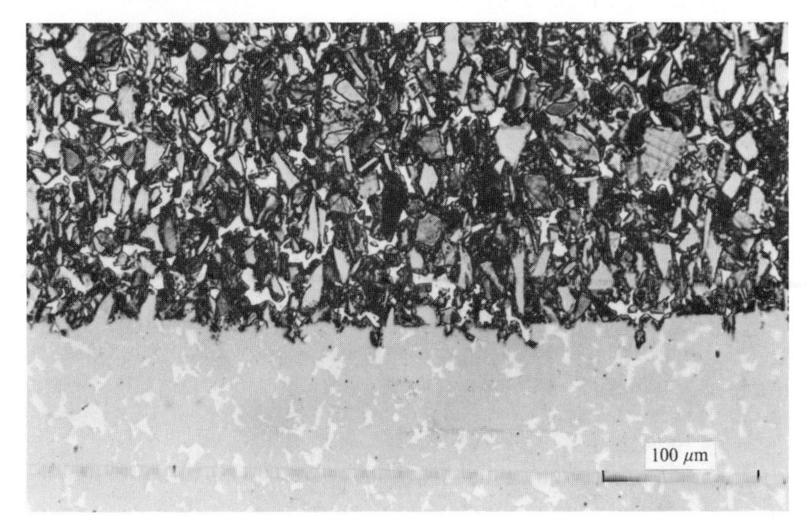

Fig. 1.7. Microstructure of self-bonded silicon carbide. The original silicon carbide grit particles can be seen in the chemically etched upper part of the specimen and the residual silicon (white in colour) in the polished lower part. (After McLaren *et al.*, 1972.)

green-forming the binders are burned out and the components then heated in contact with molten silicon. This reacts with the fine graphite to deposit more silicon carbide around the original silicon carbide particles which then bond together during the reaction as the reaction front progresses slowly through the component. The microstructure of reaction-bonded silicon carbide is shown in fig. 1.7 and comprises an intimate mixture of silicon carbide grains plus $\sim 10\%$ silicon. Both the silicon carbide and silicon phases are continuous. The necessity for the silicon in the final material is analogous to the need for porosity in the silicon nitride; it allows the reacting silicon to gain access to the centre of the component.

Reaction-bonding thus offers a wide range of green-fabrication techniques for silicon nitride and silicon carbide. A further advantage is that during the reaction-sintering process there is only a very small dimensional change, usually $<0.1\%$. This means that most of any necessary machining operations can be done in the soft green state.

1.4 Grain boundaries

In spite of the very wide range of fabrication techniques available for ceramics, the resulting microstructures possess a number of common features. One or more crystalline phases are present, sometimes with an associated glassy phase. A wide range of grain sizes is observed, generally 1–1000 μm. Porosity, which may be fine or coarse, open or closed, is also common.

Grain boundaries have a particularly profound effect on the structure-sensitive properties of ceramics such as mechanical behaviour. In pure single-phase materials, grain boundaries are simply the surfaces separating the individual crystallites, and can be regarded as thin regions of atomic disarray where the density of atoms is slightly lower than normal. Grain boundaries can thus act as sources of, and sinks for, structural defects, such as point defects and dislocations, and as sites for pores. In impure materials impurities and second phases are localised at grain boundaries. This leads to a rich variety of phenomena. At this point it is convenient to consider some energetic and geometrical aspects of grain boundaries.

When fracture occurs through a perfect single crystal two new surfaces are formed. This requires a supply of energy $2\gamma_s$, where γ_s is defined as the reversible free-energy change to create unit surface area; γ_s varies with external factors such as temperature and pressure and with internal factors such as crystal orientation (this is discussed more fully in chapter 6). The fracture process is ideally reversible in that reuniting the two fracture surfaces should result in a release of energy $2\gamma_s$. If, however, two fracture surfaces of different crystallographic orientation are brought into contact a grain boundary results. In this case the energy released is less than $2\gamma_s$ and the difference is defined as the grain-boundary energy, γ_{gb}. Under ideal conditions, therefore, the intergranular fracture energy should equal $2\gamma_s - \gamma_{gb}$ so that boundaries are more easily fractured than grains. In cases where the misorientation between the two grains is small ($<15°$), so that the boundary can be regarded as an array of dislocations, the grain-boundary energy approximates well to the total strain energy of the component dislocations. For greater misorientations, the grain-boundary energy is roughly constant.

Grain boundaries under equilibrium conditions meet the surface of the material orthogonally. Furthermore, at high temperatures, grain-boundary grooves develop along the line of intersection. The free energies of the surfaces and the boundary can be regarded as a balance of forces, as shown in fig. 1.8(*a*). Thus

$$\gamma_{gb} = 2\gamma_s \cos(\theta/2), \tag{1.2}$$

where θ is the dihedral angle. A similar situation exists at pores on grain boundaries, fig. 1.8(*b*). From measurements of grain-boundary grooves on the surface of polycrystalline alumina, Kingery (1954) found

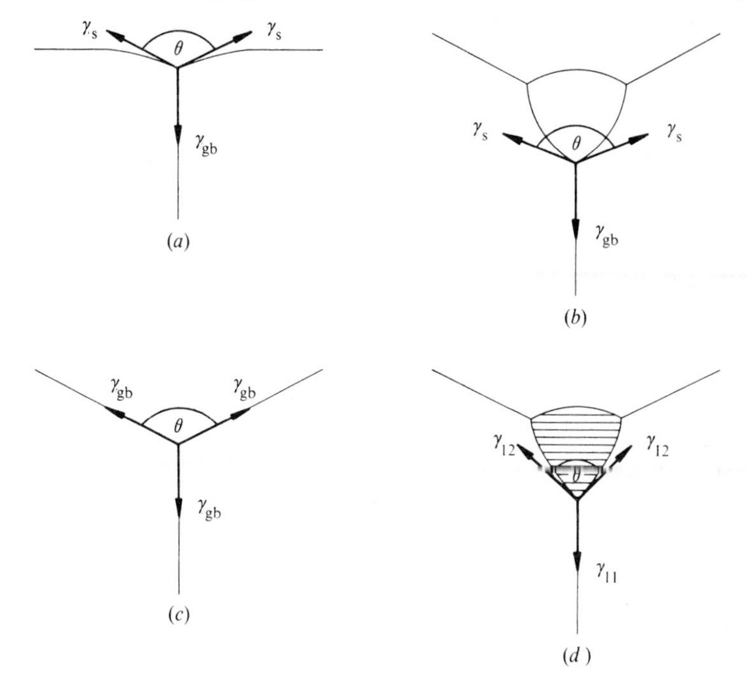

Fig. 1.8. Geometrical aspects of grain boundaries. (*a*) Two grains at a surface; (*b*) three grains at a pore; (*c*) three grains in the interior; (*d*) three grains at a second-phase particle. (After Davidge, 1976.)

Table 1.5. *Microstructural features obtained in two-phase ceramics for various ratios of boundary energies*

γ_{11}/γ_{12}	θ	Microstructure
$\geqslant 2$	0°	All grains separated by liquid phase
$\sqrt{3}$ to 2	0–60°	Penetration of all three-grain junctions
		Partial penetration between adjacent grains
1 to $\sqrt{3}$	60°–120°	Partial penetration of three-grain junctions
$\leqslant 1$	$\geqslant$120°	Isolated liquid-phase regions at four-grain junctions

$\theta = 152°$; thus $\gamma_{gb}/\gamma_s = 0.48$. An orthogonal intersection through the line joining three adjacent grains, fig. 1.8(*c*), reveals an equilibrium balance with $\theta \sim 120°$.

Similar balance-of-force arguments apply to the microstructures of two-phase ceramics, fig. 1.8(*d*). Of crucial relevance to the microstructure, and thus to the properties, is the ratio γ_{11}/γ_{12}, which controls the dihedral angle θ.

$$\gamma_{11} = 2\gamma_{12} \cos(\theta/2), \tag{1.3}$$

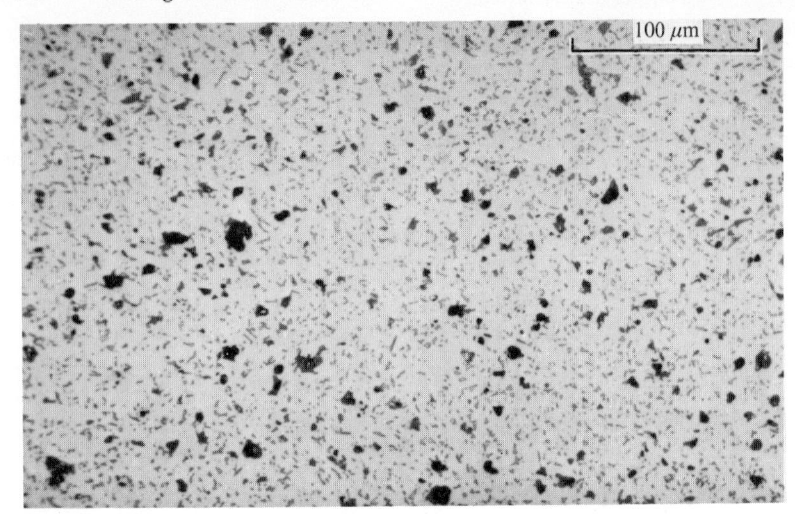

Fig. 1.9. Microstructure of 95%-alumina ceramic. The triangular, grey areas are a glassy phase and the black regions are pores.

where γ_{11} and γ_{12} are the energies of boundaries between like and unlike grains. The consequences are indicated in table 1.5.

Figures 1.9 and 1.10 show microstructures of two-phase ceramics. Figure 1.9 illustrates an alumina containing about 10% of glassy phase; θ is $< 60°$ but close to this value. The microstructure is similar to that of

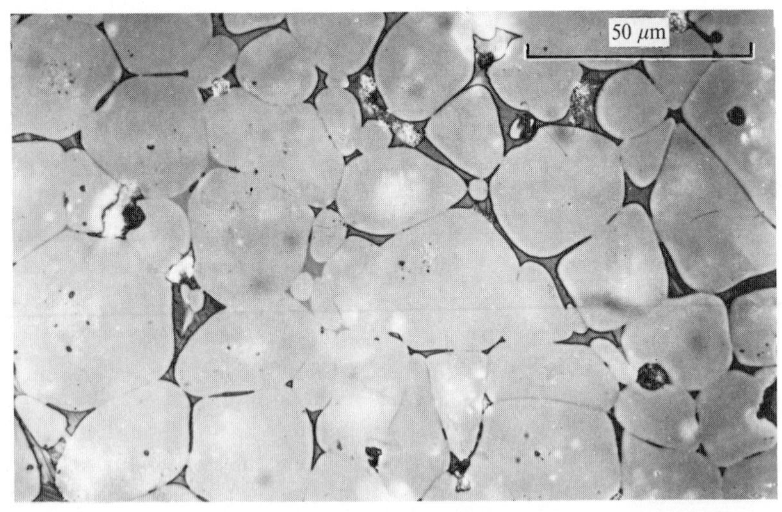

Fig. 1.10. Microstructure of magnesia refractory showing two types (light and dark) of silicate phase at the grain boundaries. Note the rounded grains compared with those in fig. 1.4. (After Davidge, 1969.)

self-bonded silicon carbide (fig. 1.7) and in both materials the minor and major phases are continuous. Figure 1.10 reveals the much more complicated microstructure of a magnesia refractory containing two different silicate phases. Here the dihedral angle is $\leqslant 60^\circ$ and there is considerable penetration of the liquid phases between adjacent grains.

In many single-phase ceramics, small but significant amounts of impurity are present; their presence may be unintentional or alternatively small amounts of impurity may have been added as a sintering aid. In these materials the impurity often tends to segregate at grain boundaries and although a discrete second phase may not be detected the effects on grain-boundary properties can be particularly marked.

In subsequent chapters we shall examine how the mechanical behaviour of ceramics is influenced by the features discussed above.

1.5 General response to stress

The above introduction serves to illustrate the extent of the field of ceramics and to define some of the more important textural features that are essential for a basic understanding of the mechanical behaviour of ceramics. For example, we shall see how elastic behaviour can be related to the atomic bonding forces, slip systems to crystal structures and fracture behaviour to microstructural features. In this section the general ways in which ceramics respond to external forces are outlined to set the general scene. No explanations are given at this stage. Emphasis is placed mainly on the effects of tensile forces because for brittle materials the tensile strength is usually about an order of magnitude lower than the compressive strength.

Stress/strain relationships

The application of external forces to a solid body results in deformation which, if it does not exceed a critical value, disappears on removal of the forces. This is elastic deformation. For a uniform rod, of cross-sectional area A, subjected to a tensile force F, the tensile stress σ is F/A and is related to Young's modulus E and the tensile strain ε by

$$\sigma = E\varepsilon. \tag{1.4}$$

When σ is increased beyond a critical value, typically of order $10^{-3}E$, there are two extreme types of behaviour: the material may fracture in a brittle manner like most ceramics, or it may deform plastically in a ductile manner like many metals. Ideal brittle fracture occurs in the absence of significant plastic flow and, after fracture, apart from the new

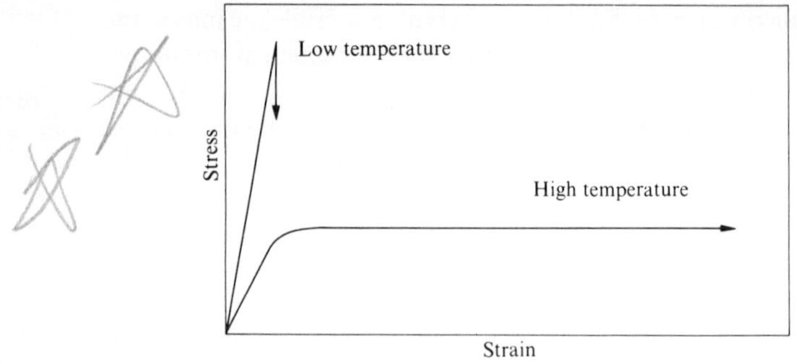

Fig. 1.11. Stress/strain curves indicating brittle behaviour at low temperature and ductile behaviour at high temperature.

fracture faces, the material returns to its original state. Plastic deformation is permanent and, unlike elastic deformation, remains after removal of the external forces.

Ceramics exhibit both types of behaviour over different temperature ranges. These are best illustrated by reference to stress/strain curves. Glass shows both extremes at particular temperatures, fig. 1.11. At low temperatures ideal brittle fracture occurs, whereas at temperatures near the softening temperature very extensive plastic flow occurs. Most materials show behaviour between these limits and the numerous deviations from the ideal will be discussed in later chapters. To indicate general trends we consider next the temperature dependence of the strength of ceramics.

Temperature dependence of strength

The tensile strength of ceramics is usually measured experimentally by three- or four-point bend tests. In the three-point bend test, fig. 1.12, the maximum stress occurs along a line on the specimen face opposite the central knife-edge. The measured fracture stress is often referred to as the modulus of rupture. The advantages of the test lie in its relative

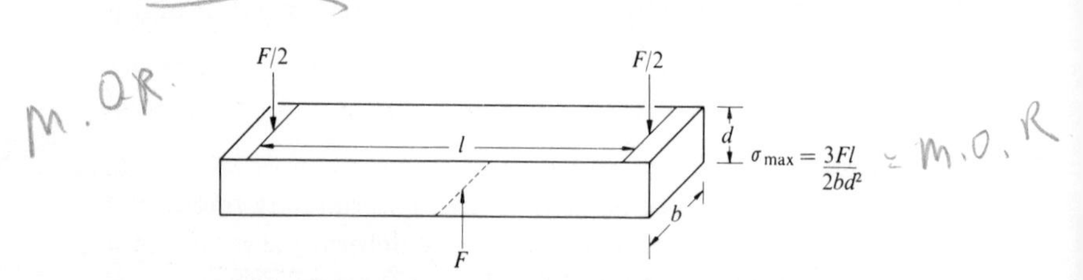

Fig. 1.12. Three-point bend test and the relationship between maximum stress, applied force and specimen dimensions.

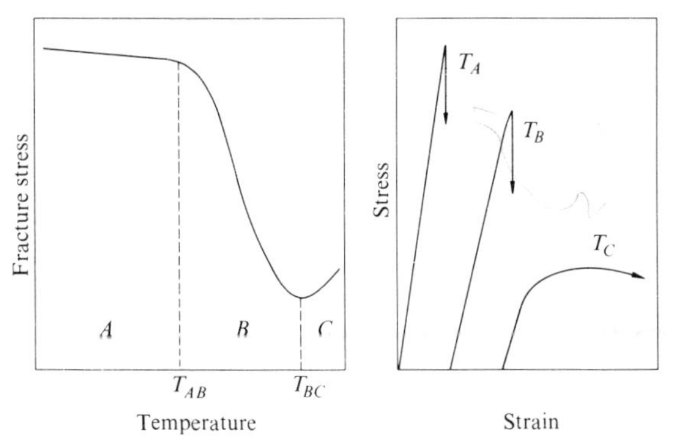

Fig. 1.13. Idealised temperature dependence of the fracture strength of ceramics, and stress/strain curves for the three temperature regions.

simplicity: the specimen has a simple form and no grips need to be attached. The disadvantages of the test are that only a small area of the specimen surface is subjected to the maximum stress and stress gradients exist both along the bar and through the thickness of the material. Fracture thus usually originates from the specimen surface but this condition relates to many service failures. Unless otherwise stated strength values quoted here refer to three-point, bend-test data.

A schematic illustration of the temperature dependence of strength for ceramics is shown in fig. 1.13. This comprises three regions labelled *A, B* and *C*. In region *A* the fracture is brittle and the fracture strain is $\sim 10^{-3}$. There is no indication of significant plastic deformation prior to failure and the strength varies little with change in temperature. In region *B* the fracture is again brittle but, in this case, evidence of slight plastic strain before fracture can be seen from the stress/strain curve. The failure strain is usually in the region 10^{-3}–10^{-2} and strength falls with increasing temperature. In region *C*, which is rarely observed except in a few single-phase polycrystalline ceramics such as MgO or UO_2, appreciable plastic flow occurs, with strains of the order of 10^{-1}, prior to failure. The critical temperatures, T_{AB} and T_{BC}, vary greatly for different ceramics. For example, T_{AB} is about 0 °C for MgO but >2000 °C for SiC. The response of ceramics to stress thus varies significantly with both type of material and variation in temperature. An important aim in later chapters is to discuss the origins of this behaviour. First, however, we consider elastic deformation.

MgO – plastic flow occurs

2 Elastic behaviour

The elastic properties of ceramics are central in determining mechanical behaviour and are directly dependent on crystal structure and bonding. Stiffness, hardness and strength are all closely interrelated, high stiffness being associated with good mechanical properties. A continuum approach is generally employed to describe elastic behaviour although we shall show how the elastic constants can be related to the atomic bonding forces for simple crystals with ionic bonding. Attention is focussed on data at ambient temperatures. The elastic moduli decrease slightly as temperature is raised and the lattice expands. A typical reduction for ceramics is ~1% per 100 K in the range 0–1000 °C. The situation is complex at high temperatures when anelastic behaviour is observed.

2.1 Elastic deformation of isotropic materials

There are few elastically isotropic materials where the strains produced by a particular stress are independent of the direction of application of the stress. Glass is an important exception. Fortunately, however, most engineering materials can be regarded as isotropic for purposes of the analysis of elastic deformation. Although individual crystallites show anisotropic behaviour and different crystal phases exhibit different elastic properties, these effects are on a much too fine scale to be of concern in engineering components. (This simplified approach cannot of course be used for highly anisotropic materials such as wood or fibre composites.)

The elastic response of isotropic materials to stress can be described by two elastic constants: Young's modulus (see (1.4)) and the shear modulus G given by

$$\tau = G\gamma, \tag{2.1}$$

where τ is the shear stress and γ is the shear strain. E and G are related through Poisson's ratio ν by

$$E = 2G(1+\nu). \tag{2.2}$$

Consider a small cubic element of material subjected to an arbitrary series of external forces. These are transmitted across each face of the cube and each may be resolved into three orthogonal components of stress, fig, 2.1. (The stresses on the faces opposite to those illustrated are equal and opposite to the indicated stresses.) The stresses σ_{xx}, σ_{yy} and σ_{zz} are tensile or compressive and the other stresses are shear stresses. In this two-suffix notation the first suffix indicates the axis along which the

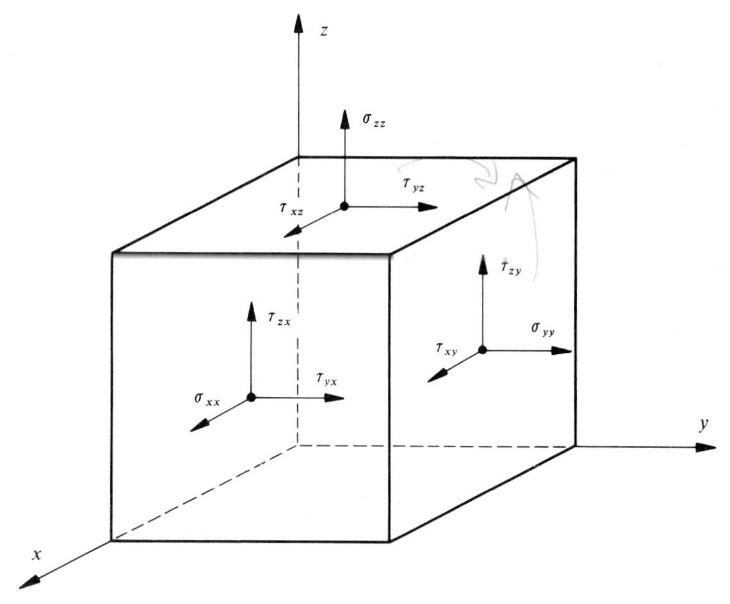

Fig. 2.1. An arbitrary series of external stresses on a unit cube of material.

stress acts, and the second suffix the axis that is normal to the plane on which the stress acts. When the cubic element is in static equilibrium it can be shown that three pairs of shear stresses must be equal; thus $\tau_{xy} = \tau_{yx}$, etc. Six components of stress are thus required to describe the arbitrary stress state of the element.

The application of a uniaxial tensile stress (σ_{xx}, say) causes an extension in the stressed direction which is accompanied by lateral contraction in the other two orthogonal directions such that

$$E\varepsilon_{yy} = E\varepsilon_{zz} = -\nu\sigma_{xx}. \qquad (2.3)$$

In a general state of stress, the tensile or compressive strain in a particular direction depends on the tensile or compressive stresses in all three orthogonal directions. The shear strains, however, are related simply to the individual shear stresses. The six strains are related to the six stresses as follows:

$$\varepsilon_{xx}=\{\sigma_{xx}-\nu(\sigma_{yy}+\sigma_{zz})\}/E$$
$$\varepsilon_{yy}=\{\sigma_{yy}-\nu(\sigma_{zz}+\sigma_{xx})\}/E$$
$$\varepsilon_{zz}=\{\sigma_{zz}-\nu(\sigma_{xx}+\sigma_{yy})\}/E$$
$$\gamma_{yz}=\tau_{yz}/G \tag{2.4}$$
$$\gamma_{zx}=\tau_{zx}/G$$
$$\gamma_{xy}=\tau_{xy}/G.$$

Finally, we note it is always possible to choose a new set of coordinate axes such that the shear stresses become zero. In this case the three remaining tensile or compressive stresses are called the *principal* stresses.

2.2 Elastic deformation of crystalline materials

For materials in the form of a single crystal we are faced with the complication that the elastic constants are anisotropic. This means that the strains produced by a particular stress depend on the direction of application of that stress. We therefore need to introduce a more complex notation. A compliance s and a stiffness c are defined by

$$\varepsilon=s\sigma$$
$$\sigma=c\varepsilon. \tag{2.5}$$

If a general stress σ_{ij} is applied to a crystal the resulting strain ε_{ij} is comprised of a number of components that are all linearly related to the stress components through the compliance constants. This means that if a cubic element of crystal is loaded in tension normal to one face then not only will it elongate in the tension direction but also it may shear such that the edges of the element are no longer orthogonal. In general there are 81 stiffness constants but considerations of symmetry reduce the number to 36 independent constants.

For isotropic materials we saw that just two independent elastic constants are required. For crystalline materials the number of constants required depends on the complexity of the crystal structure. For example, three constants are required for cubic crystal structures, five for hexagonal, nine for orthorhombic and 21 for triclinic structures. A full discussion of this problem requires introduction of the concept of fourth-rank tensors (see Nye, 1957) and this is beyond our current scope. Nevertheless a useful background can be obtained by restricting the discussion to crystals of cubic symmetry.

Elastic constants for crystals of cubic symmetry

The stiffness and compliance constants necessary for cubic crystals are

c_{11}, c_{12} and c_{44}, and s_{11}, s_{12} and s_{44}. When a normal stress is applied parallel to the cube axes of the crystal the strain in the direction of the stress is s_{11} times the stress, while the strains in the orthogonal directions are s_{12} times the stress. Poisson's ratio is therefore $-s_{12}/s_{11}$. The ratio of the shear strain to the shear stress for a stress applied parallel to one of the cube planes in a cube edge direction is s_{44}. The strains equivalent to those for the isotropic material in (2.4) are usually written thus:†

$$\begin{aligned}
\varepsilon_1 &= s_{11}\sigma_1 + s_{12}\sigma_2 + s_{12}\sigma_3 \\
\varepsilon_2 &= s_{12}\sigma_1 + s_{11}\sigma_2 + s_{12}\sigma_3 \\
\varepsilon_3 &= s_{12}\sigma_1 + s_{12}\sigma_2 + s_{11}\sigma_3 \\
\varepsilon_4 &= s_{44}\sigma_4 \\
\varepsilon_5 &= s_{44}\sigma_5 \\
\varepsilon_6 &= s_{44}\sigma_6.
\end{aligned} \tag{2.6}$$

Note that c_{ij} are *not* generally the reciprocals of s_{ij}, but are given by

$$\begin{aligned}
c_{11} &= (s_{11} + s_{12})/(s_{11} - s_{12})(s_{11} + 2s_{12}) \\
c_{12} &= s_{12}/(s_{11} - s_{12})(s_{11} + 2s_{12}) \\
c_{44} &= 1/s_{44}.
\end{aligned} \tag{2.7}$$

For isotropic materials s_{44} in (2.6) is equivalent to $2(s_{11} - s_{12})$. Thus, comparing with (2.4) shows that $s_{11} = 1/E$, $s_{12} = -\nu/E$ and $2(s_{11} - s_{12}) = 1/G$, from which (2.2) follows.

Orientation dependence of elastic constants

Both Young's modulus and the shear modulus for cubic crystals vary with the crystallographic direction $\langle hkl \rangle$:

$$\begin{aligned}
1/E_{hkl} &= s_{11} - 2\{(s_{11} - s_{12}) - \tfrac{1}{2}s_{44}\}(\alpha^{*2}\beta^{*2} + \alpha^{*2}\gamma^{*2} + \beta^{*2}\gamma^{*2}) \\
1/G_{hkl} &= s_{44} + 4\{(s_{11} - s_{12}) - \tfrac{1}{2}s_{44}\}(\alpha^{*2}\beta^{*2} + \alpha^{*2}\gamma^{*2} + \beta^{*2}\gamma^{*2}),
\end{aligned} \tag{2.8}$$

where α^*, β^* and γ^* are the cosines of the angles between the direction considered and the $\langle 100 \rangle$ axes of the cube. The term containing these

†The suffixes used above are likely to cause confusion at first sight because they are abbreviations for longer suffixes. Stresses and strains in full notation require two suffixes and stiffness and compliance constants four. The numbers 1, 2, 3 correspond to the axes x, y, z in fig. 2.1. Thus ε_1, σ_1 are abbreviations for ε_{11}, σ_{11} (or ε_{xx}, σ_{xx}); ε_4, σ_4 for γ_{23}, τ_{23}; ε_5, σ_5 for γ_{13}, τ_{13}; and ε_6, σ_6 for γ_{12}, τ_{12}. The suffixes for c and s obey a more complex scheme but for the cubic crystal discussed here the compliance constants are best visualised in terms of the simple description at the beginning of this section.

quantities is thus zero in ⟨100⟩ directions and has a maximum value of $\frac{1}{3}$ in the ⟨111⟩ directions. To illustrate the magnitude of the variations in elastic constants with orientation, tables 2.1 and 2.2 give data for magnesia single crystals. It can be seen that there is a very significant variation in the elastic moduli with crystal orientation, even for cubic crystals which have high symmetry.

Table 2.1. *Elastic constants for MgO at 25 °C (Chung, 1963)*

c_{11}	c_{12}	c_{44}	s_{11}	$-s_{12}$	s_{44}
(GN m^{-2})			(m^{-2} TN^{-1})		
289.2	88.0	154.6	4.03	0.94	6.47

Table 2.2. *Orientation dependence of Young's modulus and shear modulus for MgO at 25 °C (Chung, 1963)*

Crystal orientation	Young's modulus (GN m^{-2})	Shear modulus (GN m^{-2})
⟨100⟩	248.2	154.6
⟨110⟩	316.4	121.9
⟨111⟩	348.9	113.8

Elastic constants and interatomic forces

In principle it is possible to calculate the elastic constants from the force/distance relationships between atoms. Unfortunately, the appropriate functions are only available for the simplest crystal structures but it is nevertheless instructive to consider crystals of high symmetry in order to understand the basic concepts. The potential between two atoms in a structure follows a function of the type shown in fig. 2.2 and comprises two components: an attractive one associated with the bond type, ionic, covalent, etc. and a repulsive one associated with overlap of neighbouring electron clouds. The internal energy of the crystal U is thus the sum of these two terms. The original simple atomic model is due to Born (see Born and Huang, 1954) and this relates the elastic constants of the crystal to the internal energy function. The theory has been developed considerably since then but is still only accurate for the simplest types of bonding. In the original theory, the energy is assumed to vary according to a function of the form

$$U = -\frac{Q_1}{r^x} + \frac{Q_2}{r^y}, \tag{2.9}$$

where Q_1, Q_2, x and y are constants and r is the interatomic distance. For ionic crystals the exponent x in the attractive term is unity; the exponent y in the repulsive term is much greater, typically 6–12.

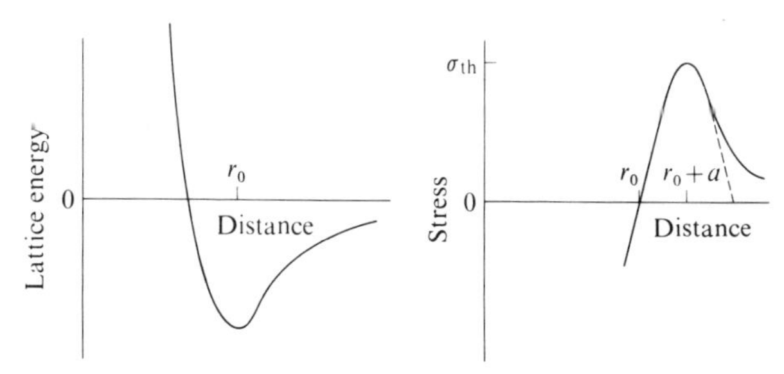

Fig. 2.2. Variation of lattice energy and stress with interatomic distance.

The elastic constants can be estimated from the function for U. A crude calculation illustrates the principles involved. The force between atoms is $(\partial U/\partial r)$ and thus the stress is

$$\sigma \sim \left(\frac{\partial U}{\partial r}\right)\frac{1}{r_0^2}, \tag{2.10}$$

i.e. the force divided by the approximate area over which it operates. Now,

$$\mathrm{d}\sigma = c\,\mathrm{d}\varepsilon = c\,\mathrm{d}r/r_0 \tag{2.11}$$

and thus

$$c = r_0\frac{\mathrm{d}\sigma}{\mathrm{d}r} \sim \frac{1}{r_0}\left(\frac{\partial^2 U}{\partial r^2}\right). \tag{2.12}$$

For a shear deformation the repulsive term in (2.9) can be neglected and thus for ionic crystals $U \propto 1/r_0$. Substition in (2.12) predicts, therefore, a proportionality between c_{44} and r_0^{-4}; this is observed experimentally, fig. 2.3, where the slope of the logarithmic plot is -4.

More sophisticated treatments show generally a very good agreement between theory and experiment for ionic crystals. For oxides the theory is less accurate and only approximate treatments are available (Wachtman, 1969).

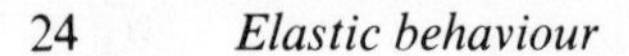

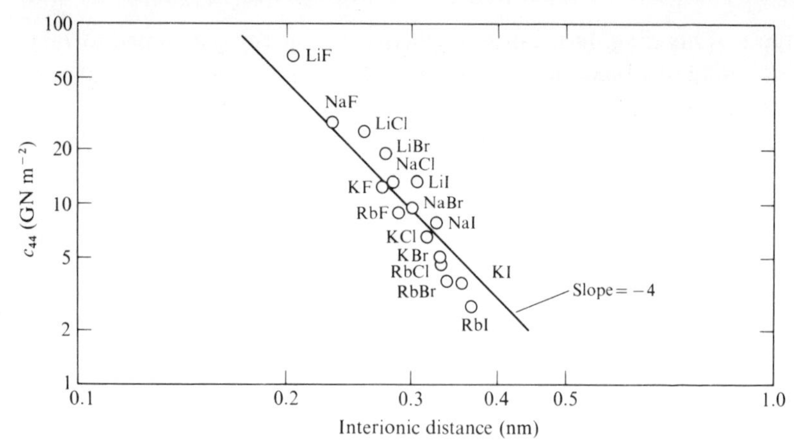

Fig. 2.3. Variation of c_{44} with interatomic distance for alkali halides. (After Gilman, 1963.)

2.3 Effects of microstructure on elastic properties

So far we have considered the elastic properties of isotropic materials and of single crystals. But most ceramics have complex microstructures comprising several phases, each of which is usually anisotropic, and also porosity. In principle, it should be possible to calculate the effective elastic constants E and G for the material from the elastic constants of the constituent phases. Alternatively, the elastic constants may be measured by direct experiments. Either way, the average elastic constants for the ceramic are known and these can then be used for calculations on engineering performance.

Although the elastic properties of the phases for a particular ceramic are usually readily defined, the elastic properties of the composite depend also on the phase geometry, which is often highly irregular. In practice, therefore, the calculation of elastic constants from basic data is not possible and only in fairly simple cases can the relevant calculations be performed. A detailed review of the recent theoretical background to the subject is given by Hashin (1968). Here, we shall consider the elastic properties of simple two-phase materials, ceramics containing pores, and polycrystals.

Simple two-phase ceramics

Let us assume for the moment that the elastic properties of each phase are isotropic. (The effects of the anisotropy of single phases is discussed below.) The elastic properties of the two-phase material must lie between those of the components. Consider a material comprising two phases with Young's moduli $E_{1,2}$ and volume fractions $V_{1,2}$. If the material is considered to have a sandwich-type construction as shown in fig. 2.4 then a stress can be applied either normal to or parallel to the component slabs. When the *stress* is applied *parallel* to the slabs the *strain* in each slab is *constant*. Young's modulus for the composite $E_{\parallel}$ is then given by the approximate formula

$$E_{\parallel} = E_1 V_1 + E_2 V_2. \tag{2.13}$$

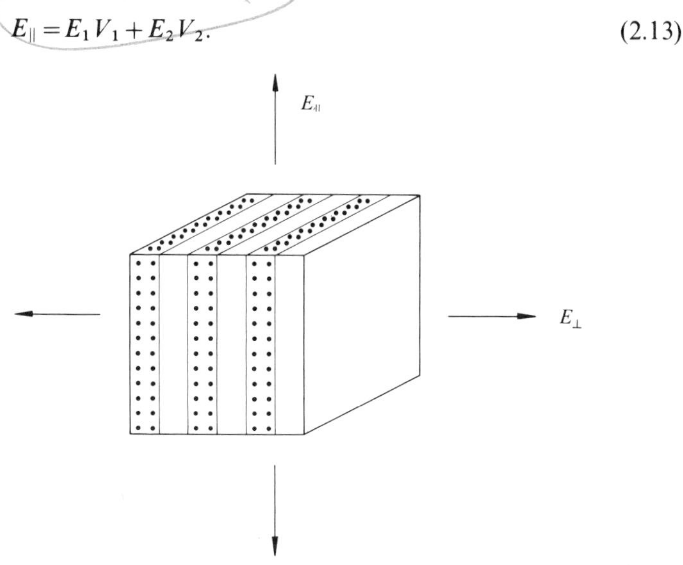

Fig. 2.4. Material of sandwich-type structure, indicating orientations parallel and perpendicular to slabs.

When the *stress* is applied *normal* to the slabs the *stress* in each slab is *constant*, and the corresponding Young's modulus $E_{\perp}$ is given by

$$E_{\perp} = \frac{E_1 E_2}{E_2 V_1 + E_1 V_2}. \tag{2.14}$$

Young's moduli of practical materials fall between these limits. Figure 2.5 shows some data of Binns (1962) for a glass containing alumina particles. The Young's moduli for the two phases are respectively 70 and 400 GN m^{-2}. The Young's modulus calculated from the parallel model gives the higher value and the experimental data for alumina particles of three sizes fall between the limits. Difficulty was found in obtaining full density of the compacts for the finest alumina particles (~ 10 μm) and thus the data points for these compacts are low (especially at 40% alumina by volume). A sensible first approximation to the elastic constant of a two-phase system is thus to take the average of the two limits.

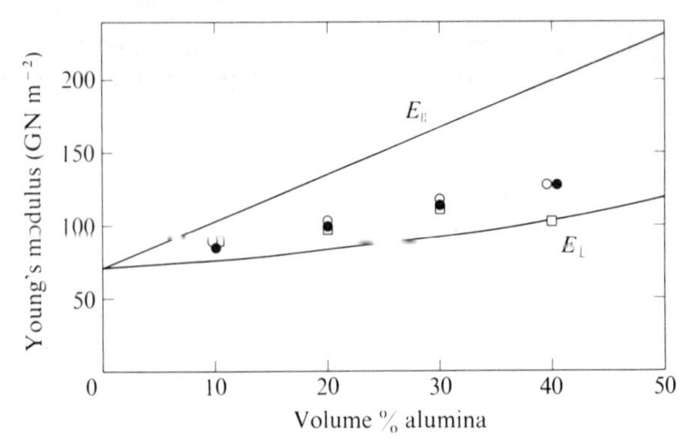

Fig. 2.5. Young's modulus of alumina/glass composites as a function of volume% alumina. ○, 180 μm mean particle size alumina; ●, 45 μm, □, 10 μm. Theoretical curves according to (2.13) and (2.14) are indicated. (After Binns, 1962.)

Polycrystalline materials

The scheme for averaging the single-crystal elastic moduli to estimate the elastic behaviour of polycrystalline materials is similar in principle to that described above. There are again two limiting cases depending on whether homogeneous strain or homogeneous stress is assumed; the original calculations were performed respectively by Voigt (1928) and Reuss (1929). Hill (1952) has demonstrated that the Voigt and Reuss estimates represent the maximum and minimum for the modulus of a polycrystalline material. Hill suggested that the arithmetic average of the two extreme values should give a good practical estimate for polycrystalline materials. Further data of Chung (1963) for polycrystalline magnesia are given in table 2.3. It can be seen that the Hill approximation gives reasonable agreement with experimental measurements.

Table 2.3. *Theoretical and experimental data for the elastic moduli of MgO at 25 °C (Chung, 1963)*

	Voigt	Reuss	Hill	Experiment
Young's modulus (GN m^{-2})	310.5	299.8	305.1	305.0
Shear modulus (GN m^{-2})	133.0	127.2	130.1	129.0
Poisson's ratio	0.167	0.178	0.173	0.18

Porosity

A porous ceramic represents the limiting case of a two-phase material, in which one phase has zero stiffness. Young's modulus decreases with increasing porosity but the rate of decrease becomes less as the porosity increases. Numerous authors have proposed equations to relate the elastic constants to the amount of porosity and most of these have been summarised by Wachtman (1969). Many equations are in the form

$$E = E_0(1 - f_1 p + f_2 p^2), \tag{2.15}$$

where E_0 is the modulus for fully dense material, f_1 and f_2 are constants, and p is the fraction porosity. The constants depend on the shape of the pores. Clearly, an array of disc-shaped pores in planes normal to the stressed direction would have a larger effect upon the modulus than an equivalent volume of spherical holes. MacKenzie (1950) has considered theoretically the situation for spherical holes. For a typical

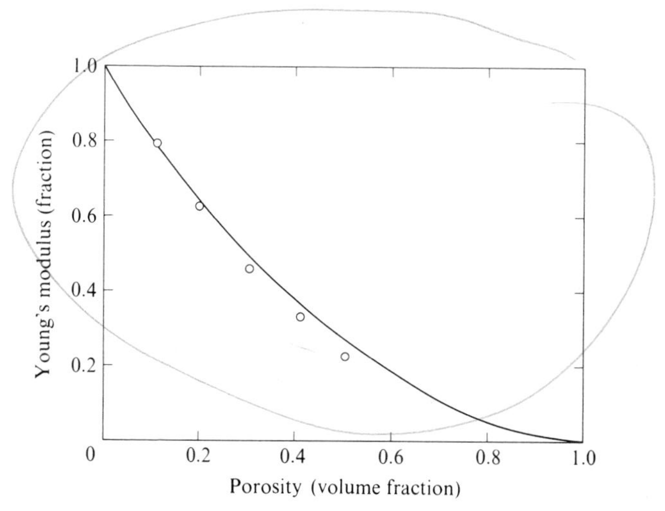

Fig. 2.6. Data of Coble and Kingery (1956) showing variation in Young's modulus with porosity in alumina, compared with theoretical estimate from (2.15).

Poisson's ratio of 0.3 the constants f_1 and f_2 in (2.15) are 1.9 and 0.9, assuming that the pores are separate in a continuous matrix. Data of Coble and Kingery (1956) are given in fig. 2.6 and the experimental measurements agree well with the theoretical predictions of MacKenzie.

The physical significance of (2.15) at low porosity, where the p^2 term is small, is that each pore has an effect equivalent to roughly twice its volume. The effect becomes less at higher porosities due to a reduced effect per pore as the pores become more closely spaced.

2.4 Estimates of the theoretical strength

Although the elastic constants can be calculated theoretically from the force/distance relationships for simple cases, estimates of the theoretical strength are less satisfactory. The relationships are more likely to be accurate near the equilibrium atomic spacing, where stress and strain are linearly related, than at large strains. Both Gilman (1963) and Kelly (1973) discuss several ways of calculating the theoretical tensile strength σ_{th}. For general purposes the rough rule of thumb that the theoretical strength is of the order 0.1 E is sufficiently accurate. It is instructive, however, to discuss the principle of a straightforward method, again based on the force/distance relationships. Zwicky (1923) used the Born relationship of (2.9) to calculate the theoretical tensile strength of sodium chloride. At the maximum stress, which Zwicky assumed to be the point of instability, $(\partial\sigma/\partial r)=0$. Recalling that $\sigma \propto (\partial U/\partial r)$, (2.10), the critical condition is thus

$$\frac{\partial^2 U}{\partial r^2}=0. \qquad (2.16)$$

Zwicky found that the theoretical tensile strength was 2 GN m^{-2} and the maximum tensile strain 0.14. Detailed calculations have also been performed for covalently bonded crystals such as diamond.

Similar considerations apply to the calculation of the theoretical shear strength of materials and details are given by Kelly (1973); the theoretical shear strength τ_{th} is approximately 0.1 G.

Kelly *et al.* (1967) give further attention to the relative values of σ_{th} and τ_{th} for a number of materials including sodium chloride, diamond and several face-centred-cubic metals. They estimated σ_{th}/τ_{th} to be 0.94 for sodium chloride, 1.16 for diamond and ~ 30 for copper, silver and gold. From considerations of the stress fields around the tips of the cracks, Kelly *et al.* produced a criterion for making a crude classification of materials on a ductile or brittle basis. They concluded that when σ_{th}/τ_{th} is large (>10) the material is consistently ductile and fails after extensive plastic flow. When σ_{th} and τ_{th} are roughly equal the material is brittle at low temperatures. There is no clear-cut boundary between these two extreme types of behaviour and when the ratio is approximately five, a large number of secondary factors need consideration.

2.5 Fracture or flow?

The theoretical strengths of materials, either in tension or in shear, are very high, but these strengths are rarely observed in practice except in certain materials in whisker or fibre form. The common strengths of ceramics are generally one or two orders of magnitude less than the theoretical values. As shown in § 1.5 the elastic region of deformation ends with the intervention either of brittle fracture or plastic flow. Fracture occurs by the propagation of cracks. The most important mechanism of plastic flow is through the motion of dislocations; other mechanisms based on the diffusion of point defects are however important at high temperatures.

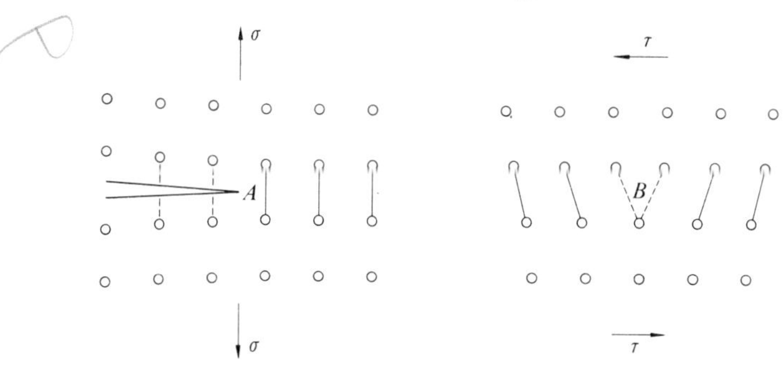

Fig. 2.7. Comparison of crack extension (from *A*) under a tensile stress, with shear at a dislocation (at *B*) under a shear stress. Solid lines indicate bonds under stress, dotted lines bonds being ruptured.

There are a number of important similarities, differences and links between crack propagation and dislocation motion. The effects of a tensile stress on a crack and of a shear stress on a dislocation are illustrated for a simple two-dimensional lattice in fig. 2.7. Both the crack front and the dislocation can be represented by lines separating the areas that have fractured or slipped from those which have not. Both the propagation of the crack and the movement of the dislocation line involve the successive breaking of lines of atomic bonds along the lines of the defects. In fracture the atomic bonds are ruptured by the passage of the crack; in dislocation flow the bonds are reunited immediately after the dislocation has passed, and the solid is thus still perfectly bonded after the passage of the dislocation. Whereas the passage of just a single crack front can lead to catastrophic failure the effect of the motion of a single dislocation is extremely small; dislocation densities of 10^{14} m^{-2} are quite common.

The relative stresses to propagate cracks or to move dislocations are obviously very important. It must be emphasised, however, that the two mechanisms are often not independent. The motion of dislocations can both form and blunt cracks; plastic flow often occurs near fracture faces as fracture proceeds.

In the next chapters we consider crack propagation, mainly from a fracture-mechanics viewpoint, and then some of the important geometrical aspects of dislocation motion. We shall then be in a position to discuss the fracture behaviour of ceramics.

3 Fundamentals of fracture mechanics

3.1 Theoretical cleavage strength and surface energy

In chapter 2 we saw how the energy function for the bonding between atoms could be used to estimate both the elastic constants of the material and the theoretical strength. By reverting to a simple model due to Orowan (1949) a good insight can be gained into the importance of surface energy when considering fracture. Orowan assumed a simple sinusoidal function relating stress to displacement (see fig. 2.2) of the form

$$\sigma = \sigma_{th} \sin \frac{\pi}{2a}(r - r_0) \tag{3.1}$$

where a is the atomic displacement corresponding to the theoretical strength. The material is assumed to be isotropic and thus (compare (2.12)) at small strains, where $(r - r_0)$ is small,

$$E = \frac{d\sigma}{d\varepsilon} = r_0 \frac{d\sigma}{dr} = r_0 \frac{\sigma_{th}\pi}{2a} \cos \frac{\pi}{2a}(r - r_0). \tag{3.2}$$

Hence

$$\sigma_{th} = \frac{2Ea}{\pi r_0}. \tag{3.3}$$

The problem reduces therefore to estimating the constant a. The key to Orowan's argument is that the work done in stressing the material to σ_{th} must at least equal the energy required to produce the two new fracture faces. In a perfect material, fracture between any pair of atomic planes is equally likely. Thus, the elastic energy in a sliver of material of thickness r_0 must be considered. Hence, if the surface energy per unit area is γ_0,

$$\int_{r_0}^{r_0+2a} \sigma dr = \frac{2Ea}{\pi r_0} \int_{r_0}^{r_0+2a} \sin \frac{\pi}{2a}(r - r_0) dr = 2\gamma_0. \tag{3.4}$$

This reduces to

$$\frac{4Ea^2}{\pi^2 r_0} = \gamma_0 \tag{3.5}$$

and eliminating a using (3.3) gives

$$\sigma_{th} = \left(\frac{E\gamma_0}{r_0}\right)^{\frac{1}{2}}. \tag{3.6}$$

High strength can thus be associated with a high surface energy and stiffness, and with a small lattice spacing. These requirements are all achieved in materials with a high density of strong bonds. Finally, recalling from § 2.4 that $a \sim 0.14\, r_0$, and substituting in (3.5) gives

$$\gamma_0 \sim \frac{Er_0}{125}. \tag{3.7}$$

Substituting values for alumina, $E = 4 \times 10^{11}$ N m^{-2}, $r_0 \sim 4 \times 10^{-10}$ m, gives $\gamma_0 \sim 1.2$ J m^{-2} which, within the accuracy of the estimates, agrees with experimental data.

The principles underlying the derivation of (3.6) are crucial in much of what follows. The essential feature is the balance between the energy required to form new fracture faces and the released elastic strain energy. This principle was used by Griffith, almost 30 years prior to Orowan, to explain the anomalously low strengths of brittle engineering materials.

3.2 The Griffith energy-balance criterion

There is a large discrepancy between the theoretical fracture strengths, $\sim 0.1E$ (estimated from (3.6) and in § 2.4), and the strengths of practical engineering ceramics (to be discussed in § 6.3) ranging typically from 0.001 to $0.002E$. Astbury (1963) analysed failure data for a much wider range of ceramics and found strength values generally in the range 0.0005 to $0.002E$; he suggested that a failure strain of 0.001 could be used as a general guide.

The key to understanding this difference between theoretical and practical strengths lies in the stress-concentrating effect of cracks. The ease with which sheet glass can be 'cut' after light scribing with a diamond tip is a powerful demonstration. 'Cut', however, is a misnomer. The function of the diamond is to introduce a shallow but *sharp* crack which can then be propagated along the sheet under the application of a small force. Note that there are two stages: *nucleation* and *propagation* of the crack.

The original analysis of the critical stress to propagate cracks was due to Griffith – the father of fracture mechanics – and was published in two classic papers (1920), (1924). The practical impetus for Griffith's work came from the observation that the surface condition (for example, filed, ground or polished) of metallic components affected the stress capability under fatigue conditions. To demonstrate the stress-raising effect of scratches, Griffith abraded soft iron wires so as to produce very fine spiral scratches at $\sim 45°$ to the wire axis. When an applied axial force exceeded a certain value ($\sim \frac{1}{4}$–$\frac{1}{3}$ of that necessary to reach the normal elastic limit) a twist, part of which remained after removal of the force, was observed in the wire. The inference was that the elastic limit had been exceeded in the vicinity of the scratches.

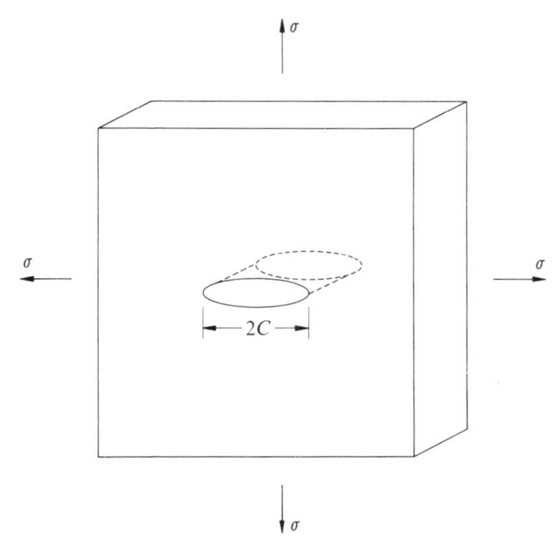

Fig. 3.1. Thin crack under biaxial tensile stress.

To explain these observations Griffith developed a new theory of rupture. From the basis that the equilibrium state of a body subjected to surface forces is one where the potential energy of the total system is a minimum, Griffith added the condition that if rupture occurs it must involve a continuous reduction in potential energy when passing from the original to the ruptured state.

The geometrical case considered is shown in fig. 3.1. A slit crack of length $2C$, approximating to a narrow ellipse, is present through the thickness of a uniform, isotropic sheet of a brittle material subjected to a biaxial stress applied in the plane of the sheet at its outer edge. (A sheet of square section is illustrated but this shape is not essential.) The stress distribution is such that at points remote from the crack the principal stresses are parallel and perpendicular to the crack, and equal. If the sheet is thin then a state of plane stress exists (zero stress normal to the sheet) whereas for thick sheets there is a state of plane strain (zero strain normal to the sheet). The equations derived for these two states, however, differ only slightly in the numerical constants.

To develop the energy balance criterion it is necessary to calculate the stress at which the crack can grow. Suppose that the stress/strain curve of the sheet containing the crack is represented by OJ, fig. 3.2. If the crack now grows in length from $2C$ to $2(C+\mathrm{d}C)$ (i.e. equal amounts at both ends) then the sheet will be less stiff. The stress/strain curve now is indicated by OKL. In practical terms one can consider this crack growth under two extreme conditions: constant stress (dead-loading) or constant strain (perfectly rigid machine). The respective strain or stress changes during the fracture increment in these two cases are JL and JK.

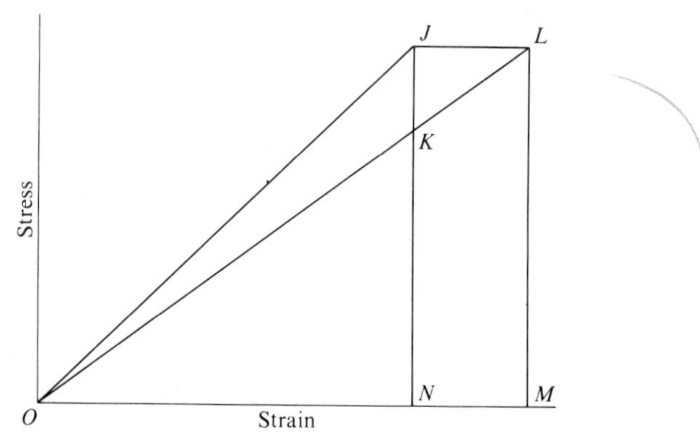

Fig. 3.2. Stress/strain curve for specimen in fig. 3.1, before crack growth (OJ) and after crack growth (OL).

Three terms need to be considered in the energy balance: (*a*) the surface energy required for the formation of new fracture faces $4\mathrm{d}C\gamma_0$ (i.e. four fracture faces of length $\mathrm{d}C$ and unit thickness), (*b*) the change in elastic strain energy in the sheet $\mathrm{d}U$, and (*c*) any external work done on the sheet $\mathrm{d}W$. The condition for fracture is thus

$$\mathrm{d}(W-U) \geqslant 4\mathrm{d}C\gamma_0. \tag{3.8}$$

The problem, therefore, is to estimate W and U.

First, however, we will demonstrate by a simple geometrical argument that the total energy available is independent of the type of loading. When fracture occurs at constant strain $\mathrm{d}W=0$, and elastic strain energy, proportional to OJK, is released ($\mathrm{d}U$ negative). When fracture occurs at constant stress $\mathrm{d}W$ is proportional to $JLMN$ and elastic strain energy increases proportional to $OLM-OJN=OJL$. In the limit, because JKL is negligible, $\mathrm{d}(W-U)$ is again proportional to OJK. Note that $dW = 2dU$; just half the external work is absorbed as elastic energy and half is available to assist crack propagation.

We can continue the discussion, relevant to both types of loading, solely in terms of U. The condition for fracture thus reduces to

$$\left(\frac{dU}{dC}\right) \geqslant 4\gamma_0. \tag{3.9}$$

Griffith calculated (dU/dC) from the results of Inglis (1913) for the distribution of stresses in a cracked plate. The presence of a thin crack in a plate reduces the strain energy by

$$\begin{aligned} U &= \pi C^2 \sigma^2 / E \quad &\text{(plane stress)}, \\ U &= \pi(1-\nu^2) C^2 \sigma^2 / E \quad &\text{(plane strain)}. \end{aligned} \tag{3.10}$$

Differentiating these equations and substituting into (3.9) thus gives for the critical condition

$$\begin{aligned} \sigma_f &= \left(\frac{2E\gamma_0}{\pi C}\right)^{\frac{1}{2}} \quad &\text{(plane stress)}, \\ \sigma_f &= \left(\frac{2E\gamma_0}{\pi(1-\nu^2)C}\right)^{\frac{1}{2}} \quad &\text{(plane strain)}. \end{aligned} \tag{3.11}$$

This discussion relates to a state of uniform biaxial tensile stress but Griffith postulated that the effect of the stress acting in the direction parallel to the slit was not significant. This was supported by his observations on the strengths of hollow glass tubes and spheres. We shall pick up this point in more detail in § 9.5.

The arguments developed so far are simply the necessary but not sufficient conditions, based on thermodynamic considerations, for cracks to extend. It has still to be demonstrated that there is also a mechanism whereby the crack can grow at these calculated stresses. For brittle fracture, the obvious additional requirement is that the stresses at the tip of the crack reach the theoretical strength and thereby permit successive breaking of the atomic bonds. The stresses around cracks are considered below.

3.3 Cracks and stress concentration

The stress-analysis work of Inglis was concerned with the stress distributions around elliptical holes. When the ellipse is infinitesimally narrow, this is a good representation of a crack. Figure 3.3 shows a half-elliptical crack of depth C situated at the surface of a body subjected to a tensile stress normal to the crack. The maximum stress concentration occurs at the tip of the crack and the stress distribution as a function of distance from the crack tip is noted in the diagram. (This stress distribution is the same as that for an internal crack of length $2C$.) Inglis showed that the stress at the tip of a narrow crack σ_m is given by

$$\sigma_m = 2\sigma(C/\rho)^{\frac{1}{2}}, \qquad (3.12)$$

where ρ is the radius of curvature at the tip of the crack. Plane-stress conditions are assumed. For fracture to occur in an ideal brittle material, a second condition is necessary in addition to the Griffith energy

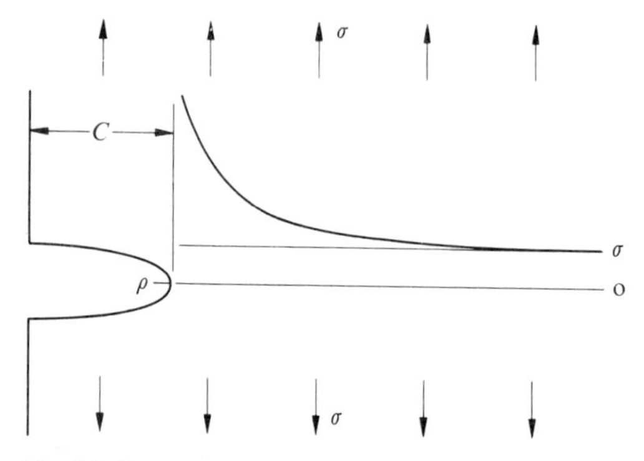

Fig. 3.3. Stress concentration at a surface flaw under a tensile stress.

requirement. The maximum stress must reach the theoretical strength of the material to provide the mechanism for fracture. Using the Orowan estimate for the theoretical strength (3.6) and equating this to σ_m gives for the failure stress

$$\sigma_f = \left(\frac{E\gamma_0}{4C}\frac{\rho}{r_0}\right)^{\frac{1}{2}}. \qquad (3.13)$$

Note from (3.12) that if the tip were assumed to have an infinitesimally small radius then the stress-concentrating effect would be infinitely large. It is clear from fig. 2.7 that a crack in a brittle material must have a crack-tip radius which is limited, because the crack passes between adjacent planes of atoms. A reasonable estimate for the tip radius is half the atomic spacing. Substitution in (3.13) thus gives for the second necessary condition

$$\sigma_f = \left(\frac{E\gamma_0}{8C}\right)^{\frac{1}{2}}. \qquad (3.14)$$

This stress is a few times lower than the Griffith stress in (3.11), so that fracture should occur at the Griffith stress. The above calculation, however, is only approximate in that it relies on Orowan's estimate for theoretical strength which is probably only accurate to within a factor of two. Nevertheless, we can conclude that for brittle materials the Griffith criterion is both a necessary and sufficient one for fracture.

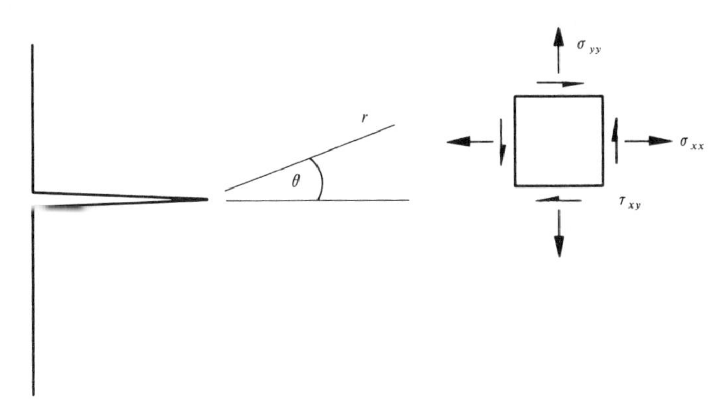

Fig. 3.4. Definition of the coordinate system for stress in the vicinity of a sharp surface flaw.

The simple ideas discussed above are the foundation for the now highly developed technical field of fracture mechanics. The main advances have occurred since around 1950 with the impetus being the requirement for sound fundamental principles on which to base the design of large engineering components. The basic Griffith concepts have thus been developed to cover both a wide range of stress distributions and loading geometries and also materials with fracture behaviour far from the ideal assumed so far. We shall consider first, developments in the techniques of stress analysis and then discuss effects caused by non-reversible fracture behaviour.

The subject of fracture mechanics is one of appreciable mathematical complexity and here we shall present only the minimum information required for our understanding of the strength of ceramics. Further details may be seen in Lawn and Wilshaw (1975). The material is assumed to behave in a linear elastic manner and we shall again consider the situation of a crack at the surface of a body, fig. 3.4. The problem is to determine stress functions to describe the stresses throughout the cracked body which satisfy the requirements of linear elasticity theory. The components of stress and strain can then be determined from the stress functions. Versatile analytical techniques have been developed for the sharp-slit case and the solutions are given below.

$$\left.\begin{matrix}\sigma_{xx}\\ \sigma_{yy}\\ \tau_{xy}\end{matrix}\right\} = \frac{K_{\mathrm{I}}}{(2\pi r)^{\frac{1}{2}}}\left\{\begin{matrix}\cos(\theta/2)[1-\sin(\theta/2)\sin(3\theta/2)]\\ \cos(\theta/2)[1+\sin(\theta/2)\sin(3\theta/2)]\\ \sin(\theta/2)\cos(\theta/2)\cos(3\theta/2)\end{matrix}\right. \tag{3.15}$$

where the coordinate system is defined in fig. 3.4. (The equations are not valid for $r \sim 0$ or for large values of $r \sim C$.) K_{I} is a constant defined as the stress-intensity factor. The suffix I refers to tensile opening mode of fracture, as relevant here (other modes with suffices II and III exist, but these involve shear fractures which are not the common failure modes of ceramics). An advantage of this analysis is the relative simplicity of the expressions for stress. Furthermore, it is possible to relate K_{I} to the Griffith failure criterion.

3.4 Relationship between surface energy and stress-intensity factor

Recourse to a simple dimensional analysis or to a more rigorous stress analysis shows that the stress-intensity factor can be related to the applied stress and the crack length by

$$K_{\mathrm{I}} = \sigma Y C^{\frac{1}{2}}, \tag{3.16}$$

where Y is a dimensionless constant which depends on the geometry of the loading and the crack configuration. Comparison with the Griffith equation (3.11) reveals the following equalities:

$$\sigma_{\mathrm{f}} = \frac{1}{Y}\left(\frac{2E\gamma_0}{C}\right)^{\frac{1}{2}} = \frac{K_{\mathrm{Ic}}}{YC^{\frac{1}{2}}}, \tag{3.17}$$

where K_{Ic} is the critical value of K_{I} for fracture.

K_{Ic} is thus equivalent to $(2E\gamma_0)^{\frac{1}{2}}$ for idealised fracture when the principles of linear elastic fracture mechanics apply. Both K_{Ic} and γ_0 can be regarded as materials constants. The preference for which parameter to use when considering fracture processes depends on several factors. Clearly γ_0 has more physical significance in that it can be visualised as surface energy and is used usually when considering strength from a materials-science viewpoint. Although K_{Ic} is a more obscure concept, it is used normally when discussing engineering design data for materials; this affords a more rational mathematical procedure. We shall use both parameters in later sections and it is important to bear in mind the close relationship between them.

3.5 Non-ideal fracture

The arguments developed so far have assumed sharp cracks in an ideal brittle material; for ceramics this is not unreasonable. Cracks are usually sharp because plastic deformation, which could in principle lead to crack blunting, is difficult. The fracture process is not far from ideal, but even in brittle materials like ceramics a number of additional energy-absorbing processes can occur during fracture. Micrographic examination of the fracture surfaces of typical polycrystalline ceramics reveals that the fracture process is quite complex. Fracture is partly transcrystalline and partly intercrystalline, and the surface is far from being planar. (It is customary to quote surface-energy values in terms of the energy per unit projected planar area of fracture face.) Even in single crystals the fracture surface is imperfect and often contains numerous steps or cleavage lines. A more serious imperfection in fracture is that there is often a thin layer adjacent to each fracture face where plastic deformation occurs. (This has been detected by etchpit techniques and by electron microscopy.) The presence of such a layer is not surprising because the stresses near the tip of the crack approach the theoretical strength of the material and thus a small zone near the crack front must be subjected to stresses in excess of the flow stress. In some materials crack branching is also observed and subsidiary cracks may form.

All of these processes consume energy in addition to the ideal fracture energy which is concerned with the breaking of atomic bonds. For practical materials we can thus define a fracture initiation energy γ_i, which is the sum of a number of terms including the ideal thermodynamic surface energy γ_0. This point will be discussed further in § 6.1. The concept of surface energy has thus been broadened to include all those energy-absorbing processes that are concomitant with fracture. It is therefore usual to refer to γ_i as an *effective* surface energy. This is a useful concept but is only valid when the effects of plastic deformation are localised near the fracture face.

3.6 Experimental techniques for the determination of fracture-mechanics parameters

The effective surface energy, and the related stress-intensity factor, are key parameters in developing our understanding of the mechanical properties of ceramics. In this section we introduce some of the more common experimental techniques that have been devised for their measurement.

The measurement of surface energy for a particular specimen geometry relies on the use of standard fracture-mechanics equations, for example (3.11) for the centre-notched plate. Measurement of the strength of such a specimen with a crack of known dimensions thus leads to a direct evaluation of the surface energy, provided that the elastic constants are known from independent measurements. The derivation of (3.11) requires evaluation of the rate of release of elastic strain energy as the crack grows. This is obtained from a detailed knowledge of the stress distribution in the vicinity of the crack and throughout the test specimen. It is possible in principle to calculate the stress distribution for any shape of specimen and examples are given later. However, a more direct experimental technique – compliance analysis – can be used, and this is presented first.

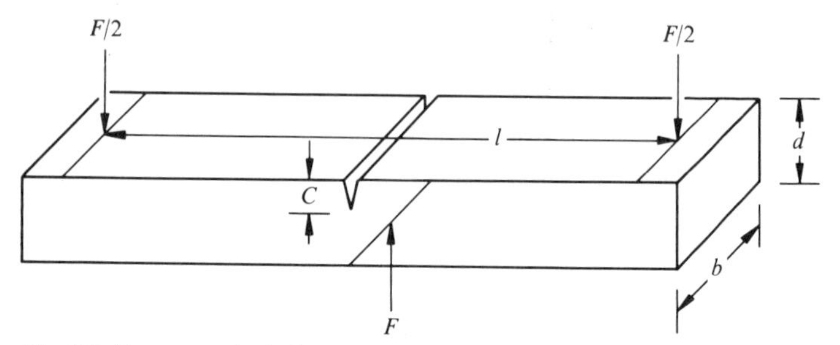

Fig. 3.5. Centre-notched, three-point bend specimen.

Compliance analysis

Although the compliance analysis method can be used for any specimen geometry, it is not often employed because a large number of specimens are required and analytical solutions are available for most specimen geometries of interest. The consideration of compliance analysis does, however, give a good insight into the fundamental principles. The method demands no knowledge of the stress distribution in the specimen and $(\mathrm{d}U/\mathrm{d}C)$ is obtained directly from the force/deflection behaviour of samples with a wide range of crack sizes.

Reference to the centre-notched, three-point bend specimen serves to illustrate the principles, fig. 3.5. (This is identical to the modulus of rupture specimen, fig. 1.12, with the addition of the central notch.) It is essential that the notch is narrow and effectively atomically sharp at the tip; we shall make this assumption for the present, but will return to this point shortly. In the original analysis of Griffith the plate had unit thickness. For practical specimens it is more usual to consider $(\mathrm{d}U/\mathrm{d}A)$ where A, the crack area, is equal to $2bC$ (fig. 3.5). The force/deflection behaviour of the specimen is of the form $F=c^*\Delta$ where c^* and Δ are the stiffness and deflection of the specimen. At the point of fracture, the stored elastic energy $U = F_f\Delta_f/2 = c^*\Delta_f^2/2$. The critical condition is $(dU/dA) = \gamma_i$, under conditions of constant deflection Δ_f, or

$$\gamma_i = -\Delta_f^2(dc^*/dA)/2. \tag{3.18}$$

Experimentally, therefore, one has to measure the specimen stiffness as a function of the crack area using a range of specimens with differing crack areas. Figure 3.6 shows how the stiffness varies with crack area (proportional to notch-to-depth ratio) for a series of glass specimens as

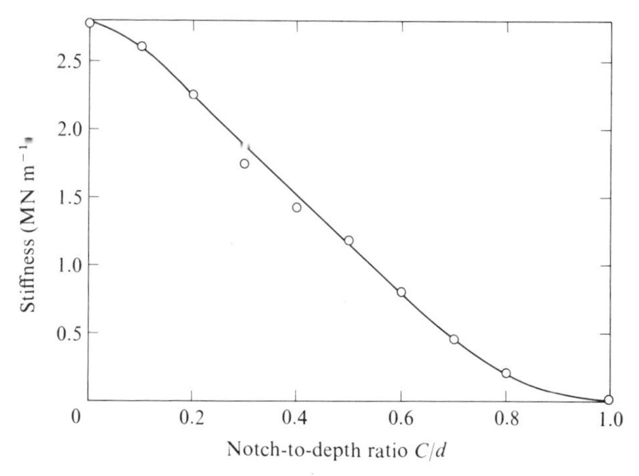

Fig. 3.6. Variation in stiffness of glass bars with depth of notch for centre-notched, three-point bend specimens. (After Davidge and Tappin, 1968*a*.)

obtained from force/deflection measurements of specimens with a wide range of initial crack sizes, at intervals of 10% of the specimen depth. It is then possible to obtain (dc^*/dA) as a function of crack area from the slope of the curve. For each notch depth the failure strain and the appropriate value of (dc^*/dA) can be inserted in (3.18) to give the effective surface energy. For example at $C/d = 0.5$: $(dc^*/dA) = 7.2 \times 10^{10}$ N m^{-3}; $\Delta_f = 14.5$ μm and hence $\gamma_i = 7.6$ J m^{-2}. This value agrees well with that determined from analytical methods.

Analytical methods

Of considerable importance when choosing suitable test geometries for ceramics is the interface requirement between the specimen and the test machine. The attachment of grips to ceramic specimens is generally difficult and leads to stress concentrations near the grips. In the bend specimen, tensile forces are generated by a compressive interaction between the machine and specimen; the bend test is by far the most commonly used method for the measurement of the strength of ceramics and is also widely used for the determination of surface energies. The more common experimental techniques in use for ceramics have been reviewed by Evans (1974*a*). Here we quote the relevant equations for the *stress-intensity factor* in terms of the applied force and other geometrical constants.

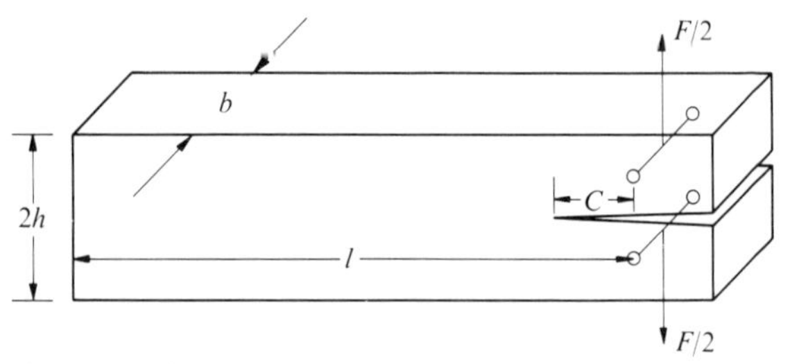

Fig. 3.7. Double-cantilever beam specimen.

Historically, the double-cantilever beam was the first specimen type to gain popularity in the testing of ceramics; this arose from measurements on single crystals each with a well-defined cleavage plane aligned along the mid-plane of the specimen. This is illustrated in fig. 3.7 and the expression for K_{I} is given by

$$K_{\mathrm{I}}=3.45\frac{FC}{bh^{\frac{3}{2}}}\left[1+0.7\frac{h}{C}\right]. \tag{3.19}$$

This specimen is still commonly used, particularly for glass. Fracture generally proceeds in a controlled manner and the dependence of K_{I} on crack velocity can be determined.

Figure 3.8 gives the calibration curves for the notched bend specimen (fig. 3.5) for a range of notch-to-depth ratios. The analyses are satisfactory for notch-to-depth ratios up to ~0.6. The factor Y is related to the stress-intensity factor by

$$K_{\mathrm{I}}=Y\frac{3FlC^{\frac{1}{2}}}{2bd^{2}}. \tag{3.20}$$

The bend specimen is particularly economical on material and can be used readily at high temperatures. It is highly recommended, therefore, when routine measurements of surface energy are required. A limitation of this specimen, however, is that once fracture is initiated the specimen almost invariably breaks into two pieces and thus only the fracture energy to initiate a crack can be estimated. This happens because the elastic, stored energy at the point of initiation of fracture is generally much greater than that required to completely fracture the specimen. This limitation is less serious in the double-cantilever beam, partly because a much larger crack surface needs to be generated to fracture the specimen.

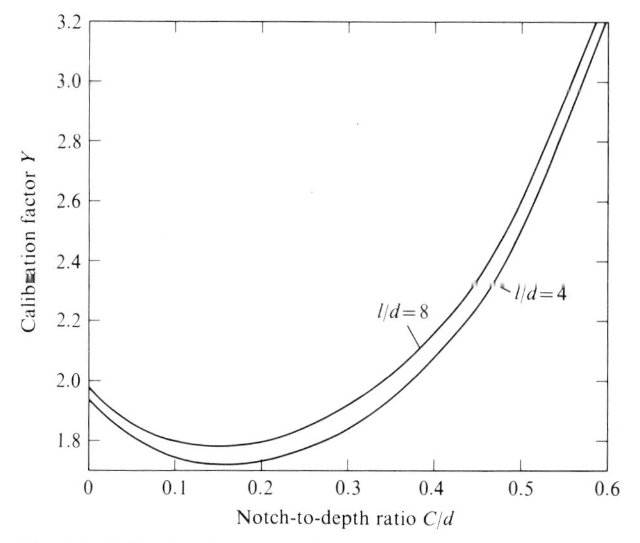

Fig. 3.8. Calibration factor Y (3.20) versus notch-to-depth ratio for centre-notched, three-point bend specimens of differing l/d ratio. (After Brown and Srawley, 1967.)

Examination of (3.19) and (3.20) shows that for the double-cantilever beam and notched beam specimens, the growth of the crack (C increasing) must be accompanied by a decreasing force F to keep the stress-intensity factor constant. The additional elastic stored energy in the loading system exacerbates this unstable tendency. Many attempts have been made, therefore, to design specimens such that the force to propagate the crack is independent of crack length and a number of speciments with this property have been developed and used for ceramics. Two modifications to the double-cantilever beam have been suggested to meet this requirement. The tapered double-cantilever beam involves the propagation of the crack into an ever increasing depth ($2h$ in fig. 3.7) of material. The constant-moment specimen is a more elegant solution and here a constant moment rather than a force is applied to a standard double-cantilever beam specimen. Neither of these specimens has been used extensively for ceramics in view of the experimental complexity.

The double-torsion geometry, where the critical force is independent of crack length and where there are no problems of attachment of the specimen to the machine, is the most satisfactory specimen developed to date. Figure 3.9 illustrates this. Under load, the two halves of the specimen rotate in opposite senses and the crack propagates along the mid-plane of the plate, often guided by a groove as shown. A peculiarity of this specimen is that, unlike all the other geometries considered, the

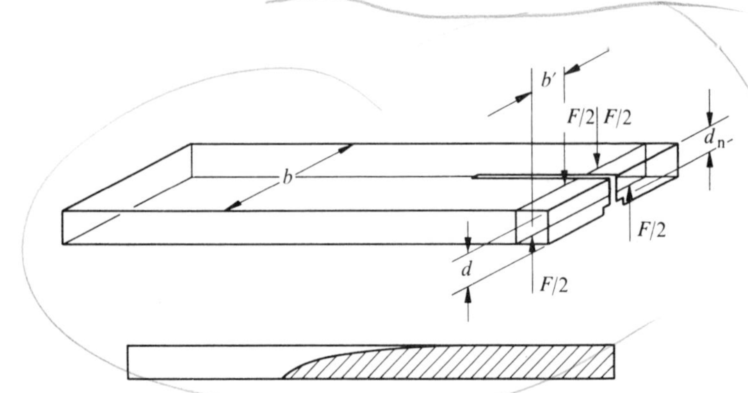

Fig. 3.9. Double-torsion specimen and crack profile.

crack front is not straight but highly curved. This means that the crack is travelling at different velocities along the crack front. In spite of this limitation the double-torsion test is widely used and is particularly suitable for studying the relationships between stress-intensity factor and crack velocity, a subject that we shall consider in more detail in § 9.2. The relevant equation is

$$K_{\rm I}=Fb'\left(\frac{3(1+\nu)}{bd^3d_{\rm n}}\right)^{\frac{1}{2}}. \tag{3.21}$$

The expressions for the effective surface energy equivalent to (3.19), (3.20) and (3.21) are obtained simply from (3.17).

Experimental requirements

The expressions quoted above assume that the principles of linear elastic fracture mechanics can be applied to the cracked specimens. This means essentially that the size of any plastic zone near the tip of the crack is sufficiently small to be negligible. A detailed analysis indicates that both the crack length and specimen thickness should be greater than about 2.5 times the quantity $(K_{\rm Ic}/\sigma_{\rm y})^2$, where $\sigma_{\rm y}$ is the yield stress of the material. Because the yield stress of ceramics is very high the ratio is quite small and this limitation is rarely critical except at high temperatures where plastic effects are pronounced. A second requirement is that the specimen dimensions should be large compared with the microstructural features of the material but this again presents little problem except in very coarse-structured materials. A third, more important, requirement is that the crack or notch should be atomically sharp at its tip. Fracture in ceramics normally occurs from small sharp cracks; to relate fracture mechanics and strength data it is thus necessary that the cracks in the fracture-mechanics specimens are similarly sharp. In many cases the ceramic machining operation to produce the groove or notch leaves sufficiently sharp cracks at the end of the notch for the data to be valid. Evans (1974*a*) has discussed various experimental arrangements for producing sharp cracks. For specimens with a long crack path such as the double-cantilever beam it is possible to load the specimen until the crack is nucleated, at which point there is a crack 'pop-in' effect, which produces the sharp crack. This pre-cracked specimen can then be used to obtain a valid result. Alternatively, crack arrest can be obtained by applying compressive stresses with vice attachments to the region ahead of the tip of the notch. Loading a wedge into the notch should result in propagation of the crack into the compressive region if the conditions are carefully controlled. A method used on notched bend specimens is to drive a wedge into the notch while the specimen sits on a firm base. A crack produced by this method, as revealed by a dye-penetrant technique, is shown in fig. 3.10. In many ceramics, results obtained from specimens with either narrow sawn notches, or sharp cracks of the same length, are identical. This, however, is not usually so for glassy materials where the force to propagate the crack from a notch is considerably greater than that for sharp crack.

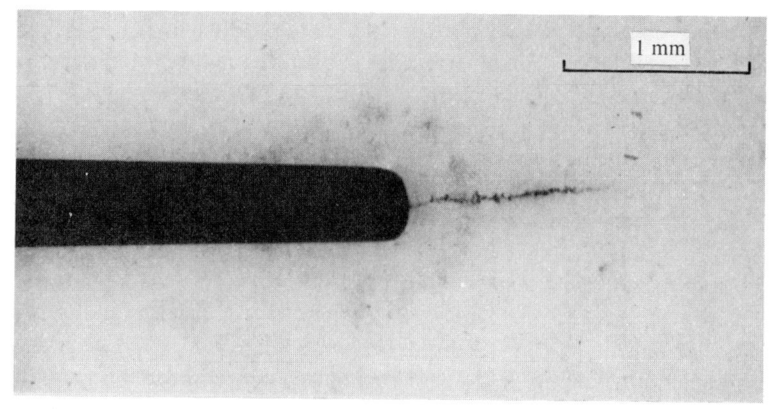

Fig. 3.10. Sharp crack (revealed by dye penetrant) at the tip of a specimen notch. (After Davidge and Tappin, 1970.)

Crack stability

Cracks in tensile specimens are usually unstable once fracture has been initiated, and fracture tends to be catastrophic. This has been considered in some detail by Berry (1960) for the sheet sample containing a central crack, fig. 3.1, but under a unidirectional stress normal to the crack. Berry's analysis shows that the locus of the critical Griffith failure criterion varies strongly with the stiffness of the specimen (controlled by the crack size) according to

$$\varepsilon_f = \frac{\sigma_f}{E} + \frac{8E\gamma_0{}^2}{A\pi\sigma_f^3}, \tag{3.22}$$

where σ_f, ε_f are the failure stresses and strains, and A is the cross-sectional area of the sheet which is assumed to be large. The form of this locus $JKLMN$ is sketched in fig. 3.11. The upper dotted line OJ, asymptotic to the curve, represents the stress/strain behaviour for a sheet with no crack. The critical fracture condition is given by the point of intersection of this locus with the stress/strain line and the locus thus separates regions of stability and instability for the crack in the sheet. The failure strain passes through a minimum value at point L when the flaw size C is given by $C^2 = A/6\pi$. Two types of behaviour are observed depending on whether the initial crack length is smaller or greater than the value at this minimum failure strain. Consider a sample with a small crack giving a stress/strain behaviour corresponding to OK. If the strain OQ on the specimen is kept constant the stress falls to the point M on the locus at the second point at which the fracture condition is satisfied for this strain. Fracture does not, however, stop at this point because the specimen contains energy, in excess of that to produce fracture surface, equivalent to the area KLM. The crack thus continues to propagate until point P is reached, where the areas KLM and MNP are equal, and the specimen has a sub-critical crack size. The specimen now contains elastic energy equivalent to area OPQ; surface energy equivalent to area $OKLMP$ has been absorbed during this process. For specimens with a large initial crack, say represented by OM, the crack is stable when the fracture strain is reached and propagates only if further work is done on the specimen. Then the stress/strain coordinates follow the locus curve towards N.

Confirmation of this idealised behaviour can readily be obtained experimentally and fig. 3.12 shows three force/deflection curves for bars of alumina deformed in three-point bending having notches 0.1, 0.2 and 0.7 of the specimen depth. For specimens with very shallow notches, it is

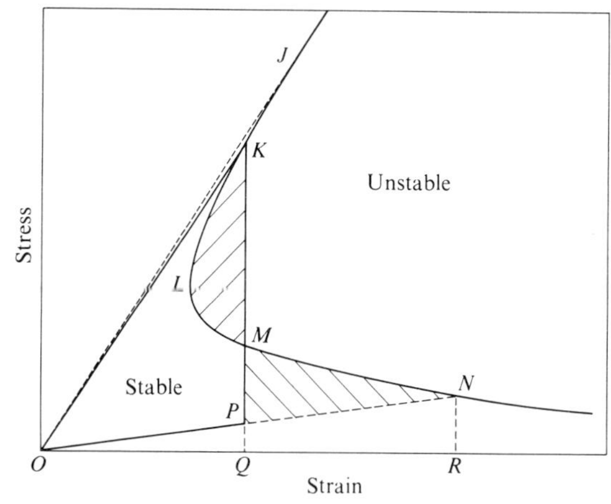

Fig. 3.11. Locus of the Griffith failure criterion (*JKLMN*) for specimens with a range of flaw sizes. (For additional discussion see § 8.4.) (After Berry, 1960.)

not possible to retain the crack in the specimen because the situation is complicated by the elastic stored energy in the testing machine. This stored energy increases the tendency to overshoot the equilibrium point. For specimens with larger cracks the crack is retained in the specimen after fracture has been initiated and complete failure is only obtained on doing further work on the specimen. For this particular specimen geometry and material, completely controlled crack propagation is obtained when the notch-to-depth ratio is >0.6.

Specimens exhibiting controlled or semi-controlled failure characteristics have been used to determine the energy for fracture propagation in

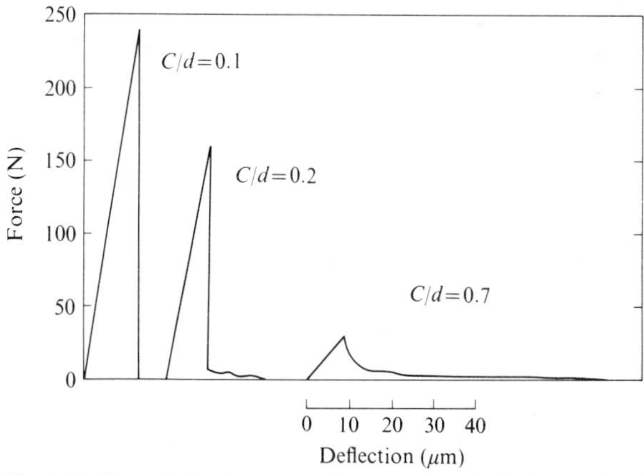

Fig. 3.12. Force/deflection curves for centre-notched, three-point bend specimens of alumina with three notch-to-depth ratios.

ceramics. This quantity is usually referred to as the work of fracture γ_f. The original geometry was a three-point bend specimen with a notch introduced by two saw-cuts leaving a triangular cross-section at the point of maximum load. The apex of the triangle was subjected to the maximum load so that the crack would move into an ever-widening region of material as it propagated through the specimen, to give a greater degree of control of crack propagation. Similar results may be obtained with the more convenient centre-notched bar, fig. 3.5. The work of fracture is defined simply as the work done on the specimen during the complete fracture process (indicated by the area under the force/deflection curve to complete failure) divided by twice the cross-sectional area at the notched part of the specimen. The advantage of this technique is that it requires no knowledge of the stress distribution in the specimen or the elastic properties of the material being tested. The work-of-fracture test has been used particularly in comparing the energies for fracture propagation in ceramics with those of other classes of materials such as metals, composites and plastics. This is a useful test for such purposes in that the fracture energies of these materials are widely different. The major disadvantage of the work-of-fracture test is that it integrates the total fracture-energy requirement over the complete fracture process. From fig. 3.12 we can infer that fracture occurs over a wide range of crack velocities particularly when shallow notches are used. One would therefore expect the work of fracture to vary with the depth of notch and, furthermore, that the work of fracture should not necessarily equal the effective surface energy for fracture initiation. Figure 3.13 confirms this, showing data for fracture-initiation energies and work of fracture as a function of notch-to-depth ratio for bars of alumina. Whereas γ_i is essentially constant with varying notch-to-depth ratio, γ_f falls rapidly as the notch-to-depth ratio increases. This indicates that rapidly moving cracks absorb more energy than do slowly moving cracks. For a slow-crack propagation in the work-of-fracture test at large notch-to-depth ratios, γ_f is close to γ_i.

3.7 The stress to propagate circular and elliptical flaws

For the specimen types described above the cracks have been characterised by a single parameter – their depth or length. The cracks, which are generated from a machined slit or notch, are usually through the thickness of the specimen and the crack front is linear. (The most important exception to this is the curved front of the crack in the double-torsion specimen.) The flaws from which fracture originates in un-notched ceramic specimens are not of this type. As we shall see later, they are usually associated with particular microstructural features such as grains or pores and, therefore, tend to approximate more to a semicircle or circle in form. Flaws are often sited at the specimen surface.

The general fracture equation (3.17) needs modification to account for flaw shape by introduction of a constant Z such that

$$\sigma_f = \frac{Z}{Y}\left(\frac{2E\gamma_i}{C}\right)^{\frac{1}{2}}. \tag{3.23}$$

In the discussion that follows we shall assume that the flaw is situated in a uniform tensile stress. This is appropriate to the bend test, where the

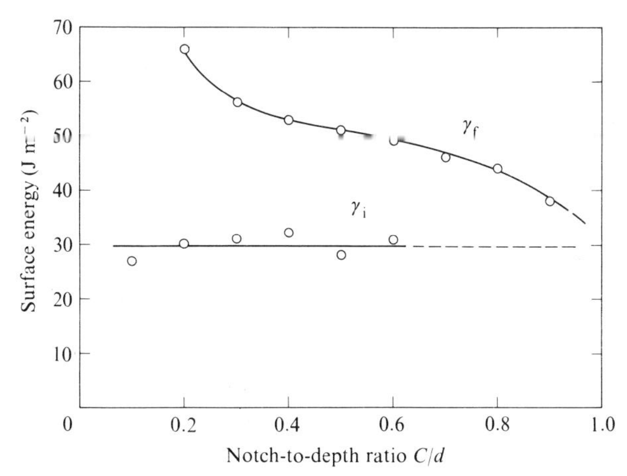

Fig. 3.13. Variation of γ_i and γ_f with notch-to-depth ratio for alumina. (After Davidge and Tappin, 1968*a*.)

tensile stress is sensibly constant over the dimensions of the flaw which is small compared with the specimen dimension. Sack (1946) considered the simplest case for an internal circular flaw, equivalent to a semicircular surface flaw, and showed that $Z=\pi/2$. The strength of a specimen with such a flaw is thus 1.57 times greater than the equivalent specimen with a through-the-thickness crack ($Z=1$).

Discussion of the more general case for a semi-elliptical surface flaw (fig. 3.14), with β/α varying, has been given by Evans and Tappin (1972) and by Bansal (1976). The variation of Z with β/α is shown in fig. 3.14. Bansal demonstrated that for most elliptical flaws of practical significance, where $0.2<\beta/\alpha<3.0$, the data in fig. 3.14 can be approximated by $Z^2=2.82\beta A_c^{\frac{1}{2}}$, where A_c, the area of the flaw, is $\pi\alpha\beta/4$. Substitution into (3.23) thus gives

$$\sigma_f \sim \frac{1.68}{Y}\frac{(2E\gamma_i)^{\frac{1}{2}}}{A_c^{\frac{1}{4}}}. \tag{3.24}$$

The error in using this approximation is $<5\%$ for the range of ellipticity stated and can usefully be applied for flaws which are not truly elliptical. The stress-intensity factor is constant for all points on the crack front only for semicircular cracks. For other cases of ellipticity the variation in stress-intensity factor along the crack front tends to convert the crack front to a semicircular shape as the crack propagates.

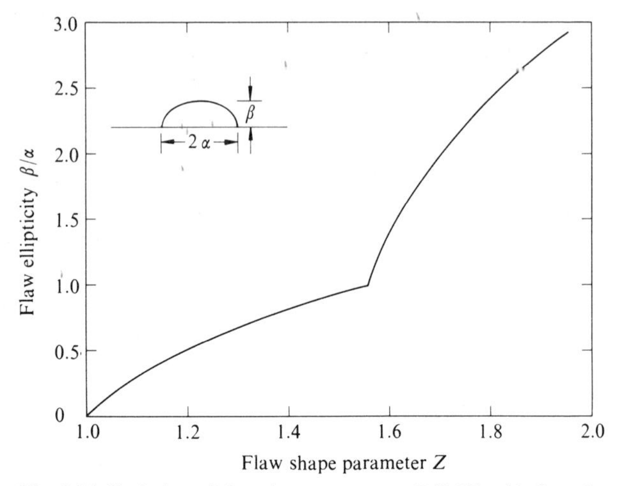

Fig. 3.14. Variation of flaw shape parameter Z (3.22) with flaw shape. (After Randall, 1967.)

When the spacing between neighbouring flaws is comparable to their size they may link together at a stress lower than that to propagate a single flaw. The calculations of Paris and Sih (1965) show that the factor Z in (3.23) for an array of flaws, of length $2C$ and spacing between centres $2S$, through the thickness of a specimen under a tensile stress (as in fig. 3.1) is

$$Z=\left\{\frac{2S}{\pi C}\tan\left(\frac{\pi C}{2S}\right)\right\}^{-\frac{1}{2}}. \tag{3.25}$$

Z thus tends to 1 at large values of $2S$ and only becomes significantly less than 1 when the flaws are close together. For example, at $2S=4C$, $Z=0.89$. For an array of circular flaws Z in (3.25) is increased by $\sim\pi/2$, which is the same ratio as that for single flaws.

4 Geometrical aspects of dislocations

Dislocation motion is the most important mechanism of plastic deformation in crystalline materials. Dislocations are line defects which can occur in any crystal lattice and almost invariably do – the dislocation-free crystal is still a rarity. As we have seen, most ceramics apart from glasses are comprised of crystalline phases although many also contain a glassy phase, often situated at grain boundaries. Studies of dislocations in ionic and ceramic crystals, particularly in single-crystal form, have played a significant role in developing understanding of dislocation theory, but this is not our concern here.

The aim in this chapter is to emphasise how the geometrical aspects of dislocations in ceramics are different from those in metals and to show why even pure single-phase ceramic polycrystals are usually brittle. Our interest lies mainly in explaining why dislocation activity in ceramics is restricted and the conclusions are of great importance; by understanding the inhibited behaviour of dislocations we can see why brittle failure is the norm.

4.1 Slip systems

The dislocation is a mechanism for shearing between adjacent atomic planes by the successive breaking of bonds along the line of dislocation. The amount of shear produced by the dislocation is characterised by its Burgers vector $\boldsymbol{b}$. The elastic strain energy associated with the dislocation is proportional to b^2. For each crystalline lattice, therefore, it is generally the dislocations with the shortest Burgers vector that form; this is in accordance with the observation that slip occurs on close-packed planes in close-packed directions.

The operative slip systems for a particular material depend primarily on its crystal structure, although a number of other factors also need consideration. It is convenient first to consider the principles with reference to materials of the NaCl-structure because of its simplicity and the large amount of available data. This structure is illustrated in fig. 1.3(*b*) and it typified by the ceramic MgO. The Burgers vector is the distance between like ions in the structure and thus the shortest vector is in the $\langle 110 \rangle$ direction. This is, in fact, the observed vector but it occurs in several crystal planes such as $\{100\}$, $\{110\}$ and $\{111\}$. In many crystal structures the active slip planes are the most closely packed and the most widely spaced. If h is the spacing between adjacent slip planes the shear angle is $\tan^{-1}(b/h)$, and this led Chalmers and Martius (1952) to postulate that b/h should be a minimum for the plane of easiest gliding. For the three slip planes b/h increases in the order mentioned above, yet the observed slip plane in MgO is the {110} plane and not the {100} plane as expected. To understand this we need to consider the relative positions of the ions when shear occurs on the various planes. Figure 4.1

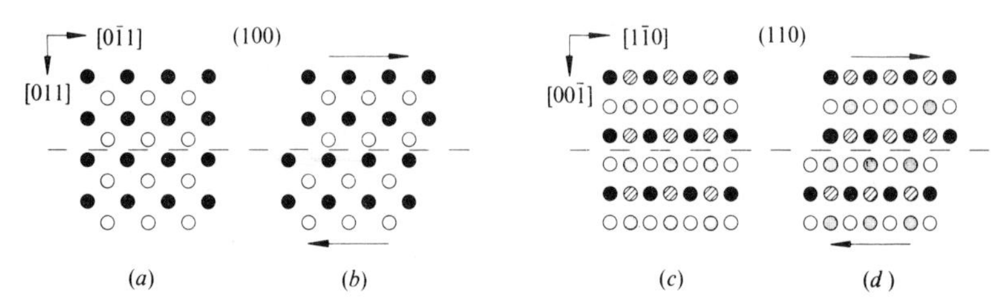

Fig. 4.1. Atomic positions for equilibrium and maximum shear positions in a NaCl-type lattice. Open/filled circles: Mg/O in plane of diagram. Cross-hatched/dot-tinted circles: Mg/O in planes $\frac{1}{2}b$ above and below plane of diagram. (*a*) and (*b*): {110}⟨1$\bar{1}$0⟩ shear. (*c*) and (*d*): {100}⟨1$\bar{1}$0⟩ shear.

illustrates this for shear on the {110} and {100} planes, where the ionic positions are shown for the undisturbed lattice and also for a displacement of one-half the Burgers vector where the lattice disturbance is a maximum. (It will assist in following the argument to use as an aid a three-dimensional ball model of the structure.) For {110} slip, ions of like charge are not brought into juxtaposition during glide. This is not so for {100} slip. Each ion at the mid-shear position is equidistant from equal numbers of nearest-neighbour and next-nearest-neighbour ions of each charge across the slip plane, and this results in no net binding; {110} glide is thus favoured in MgO.

Most crystals with NaCl-structure slip on {110}⟨1$\bar{1}$0⟩ systems but this is not universal. The observed slip systems for a wide range of NaCl-type crystals have been discussed by Gilman (1959) who noted that two crystals, PbS and PbTe, were found to slip on {100} planes. This anomaly was first highlighted by Buerger (1930) who pointed out that the tendency for slip on {100} planes increased as the polarisability of the ions increased, and table 4.1 illustrates this. Polarisability increases as the size of the ion increases and is a measure of the ease of deformability of the ion. The electrostatic faulting for {100} slip can thus be more readily accommodated in crystals having ions of high polarisability.

A related effect is that of increasing the temperature. Those crystals which slip on {110} planes are often observed to slip additionally on {100} planes at high temperatures where the lattice is more relaxed.

We find then that in crystals with ionic bonding the operative slip systems depend not only on the crystal structure, but also on the ionic positions in the dislocation core during shear, with particular reference to the polarisability of the ions.

Table 4.1. *Comparison of the primary glide planes of NaCl-structure crystals with ionic polarisability and lattice constant. (After Gilman, 1959)*

Crystal	Primary glide plane	Polarisability (10^{-30} m^{-3}) Anion	Cation	Total	Lattice constant (nm)
LiF	{110}	0.03	1.0	1.0	0.401
MgO	{110}	0.09	3.1	3.2	0.420
NaCl	{110}	0.18	3.7	3.9	0.563
PbS	{100}	3.1	10.2	13.3	0.597
PbTe	{100}	3.1	14.0	17.1	0.634

Similar, but more geometrically involved, considerations apply to more complex crystal structures and we next discuss alumina. This has a hexagonal structure and is illustrated in fig. 1.3(*c*). Dislocation geometry for alumina has been discussed in detail by Kronberg (1957). The unit cell for the alumina lattice is made up of six layers of ions parallel to the basal planes which are comprised alternately of aluminium and oxygen ions. All sites are filled on the oxygen planes but on the aluminium planes only two-thirds of the sites are occupied. The unoccupied sites, or holes, are arranged regularly, but differently, on each of the three aluminium planes in the unit cell. Figure 4.2 illustrates two of the adjacent basal planes; a completely filled oxygen plane plus a partly filled aluminium plane, the aluminium ions lying in the interstices between the oxygen ions. Note that both the aluminium ions and holes define a regular hexagonal mesh.

Alumina does not exhibit dislocation mobility at ambient temperatures, a point that we return to later. However, deformation of single crystals at temperatures of ~ 1300 °C or higher shows that appreciable slip can occur on the {0001} basal plane. This is the closest-packed plane of alumina and according to the principles outlined above should be the expected slip plane. The detected slip direction is $\langle 11\bar{2}0 \rangle$ which, referring to fig. 4.2, is at 30° to the close-packed direction of the oxygen ions.

This appears to contradict the general principles laid down so far but this anomaly is readily resolved by taking cognisance of the partly filled layers of aluminium ions. The geometrical patterns of both the aluminium atoms and the holes must be restored after shear. Reference to the holes in fig. 4.2 shows that for the $\langle 10\bar{1}0 \rangle$ direction the repeat distance for holes is $\sqrt{3}$ times greater than that in the $\langle 11\bar{2}0 \rangle$ direction.

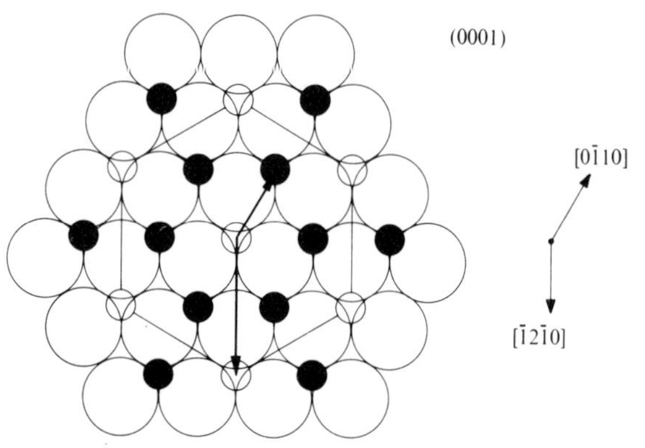

Fig. 4.2. Atomic arrangements in basal planes of α-Al_2O_3. Large circles: O; small filled circles: Al; small open circles: vacant Al sites. Repeat distance for oxygen atoms in $\langle 10\bar{1}0 \rangle$ equals length of short arrow. Repeat distance for aluminium holes in $\langle 1010 \rangle$ equals three times length of short arrow, and in $\langle 11\bar{2}0 \rangle$ the length of long arrow. (After Kronberg, 1957.) (See also fig. 1.3 (*c*).

On the dislocation-energy argument, therefore, the energy of dislocations with the former Burgers vector would be three times that of those with the latter (observed) Burgers vector. Because each atomic plane adjacent to the slip plane comprises just one type of ion there are no problems of juxtaposition of like ions in this structure, in contrast to MgO. There is thus excellent agreement between the expected and observed slip systems for alumina.

The slip systems for a wide range of ceramics have been reviewed by Davidge (1973), Chin (1975) and Evans and Langdon (1976); the observed systems for a number of common ceramic crystals are summarised in table 4.2. This includes the primary (most common) and secondary (next most common) system, and the temperature at which slip is first observed on these systems. The other information in the table will be commented upon later.

4.2 Plasticity of single crystals

Magnesia is the ceramic that shows most pronounced plasticity at ambient temperatures. MgO single crystals can be cleaved readily on

Table 4.2. *Primary and secondary slip systems of ceramics. (After Davidge 1973; Chin 1975; Evans and Langdon 1976)*

Material	Crystal structure	Slip systems		No. of associated independent slip systems		Temperature for appreciable slip (°C)	
		Primary	Secondary	Primary	Secondary	Primary	Secondary
Al_2O_3	Hexagonal	$\{0001\}\langle 11\bar{2}0\rangle$	Several	2		1200	
BeO	Hexagonal	$\{0001\}\langle 11\bar{2}0\rangle$	Several	2		1000	
MgO	Cubic (NaCl)	$\{110\}\langle 1\bar{1}0\rangle$	$\{001\}\langle 1\bar{1}0\rangle$	2	3	0	1700
$MgO.Al_2O_3$	Cubic (spinel)	$\{111\}\langle 1\bar{1}0\rangle$		5		1650	
β-SiC	Cubic (ZnS)	$\{111\}\langle 1\bar{1}0\rangle$		5		>2000	
β-Si_3N_4	Hexagonal	$\{10\bar{1}0\}\langle 0001\rangle$		2		>1800	
TiC	Cubic (NaCl)	$\{111\}\langle 1\bar{1}0\rangle$		5		900	
UO_2	Cubic (CaF_2)	$\{001\}\langle 1\bar{1}0\rangle$	$\{110\}\langle 1\bar{1}0\rangle$	3	2	700	1200
ZrB_2	Hexagonal	$\{0001\}\langle 11\bar{2}0\rangle$		2		2100	

{100} planes and regular parallelepiped specimens for tension, compression or bending tests are easily produced. During these tests equal stresses are produced on four of the slip systems, the remaining two being unstressed. The extent of plasticity is determined partly by the slip density on each system, and partly by the presence or absence of microcracks on the surface of the specimens.

As-grown crystals contain a low density of inherent dislocations but these are in random orientations and usually pinned by impurity precipitates. High stresses are thus required to move these dislocations. The cleavage process also introduces fresh dislocations on the conventional glide planes which can propagate at modest stresses. Dislocations can multiply under stress by a cross-glide mechanism (Johnston and Gilman, 1960) so that a single dislocation loop can move across the crystal and form a glide band with a high dislocation density which can then broaden sideways throughout the crystal. A high density of slip sources usually leads to interactions between dislocations on different slip systems thus limiting the amount of plasticity. On the other hand, when dislocation activity stems from a few sources on *parallel* planes extensive plastic deformation is possible ($\sim 10\%$).

As-cleaved specimens generally also contain a number of small microcracks at the specimen surface as a result of imperfect cleavage. Brittle or ductile behaviour is thus determined initially by the relative stresses to propagate these cracks, according to the Griffith equation, and the dislocation flow stress of the crystal (typically ~ 100 MN m^{-2}). In some cases the stress to propagate the cracks is the lower and fracture thus occurs in a brittle manner with very little plasticity ($< 0.1\%$). In other cases the microcracks are only a few μm in size and are about an order of magnitude less than the critical Griffith size. A simple calculation illustrates this. The stress to propagate cracks of semicircular shape in a bend specimen is, from (3.23) and fig. 3.8, $\sigma_f = \pi (2E\gamma_i/C)^{\frac{1}{2}}/4$. Substituting $\gamma_i = 1$ J m^{-2}, $E = 300$ GN m^{-2} and the flow stress, 100 MN m^{-2}, gives a critical flaw depth $C = 37$ μm. Clarke *et al.* (1962) observed that cracks of sub-critical size would grow slowly, accompanied by plastic deformation, with fracture occurring after small plastic strains when the crack had grown to the critical Griffith size. They postulated a dislocation model for this which is illustrated in fig. 4.3. The crack is situated near a corner of the crystal and, under stress, dislocations are nucleated by the enhanced stress at the tip of the crack. Although the initial dislocations tend to move away from the crack the resulting glide band broadens by the cross-slip mechanism. Other dislocation sources on parallel planes then operate and send dislocations of the opposite geometrical sign back towards the crack-tip region. Dislocations entering the crack-tip region supplement the strain energy near the tip of the crack and this favours a slow growth of the crack as shown. Essentially, this mechanism can be regarded as reducing the effective surface energy for crack propagation, the moving crack gaining energy on passing through the strain fields of dislocations sited near its path.

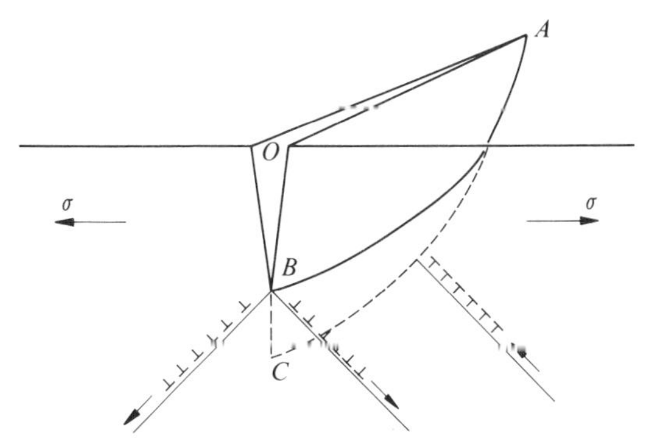

Fig. 4.3. Model for microcrack growth assisted by dislocation motion. (After Clarke *et al.*, 1962.)

The rather random behaviour of as-cleaved crystals led researchers in this field to polish their crystals prior to deformation, thereby eliminating the microcracks produced during cleavage. Now we shall be concerned with both the nucleation and propagation of cracks. In this case the deformation is controlled by the distribution of slip throughout the crystal. Figure 4.4 shows some tensile stress/strain curves on polished MgO single crystals obtained by Stokes *et al.* (1961). When slip occurs on just a few intersecting slip bands only a small amount of plasticity is observed. When there are many intersecting slip bands a much greater amount of plasticity occurs, but the greatest plasticity is obtained in crystals which deform by just a single expanding slip band. In all cases the final fracture is of a brittle type and this leads to the conclusion that plastic deformation is responsible for the nucleation of new crack sources. To understand these observations we need to discuss how cracks may be generated by dislocation movements.

A number of mechanisms have been proposed for the generation of cracks following dislocation motion, but the most important of these involves the pile-up of dislocations at some barrier in the material. This barrier may take several forms including slip bands of other slip systems, impurity precipitates or grain boundaries. For single crystals the most important barrier is the slip band of a slip plane other than that of the dislocations considered.

The basic geometrical aspects of the mechanism are illustrated in fig. 4.5, due originally to Zener (1948). Operation of a slip source S under a stress results in dislocations being forced towards the barrier. As the stress is increased the leading dislocations are forced closer together until, at a sufficiently high stress, they coalesce, resulting in a micro-crack, as indicated. To obtain the crack-nucleation stress we need to equate the stress near the pile-up of dislocations to the cohesive strength of the crystal.

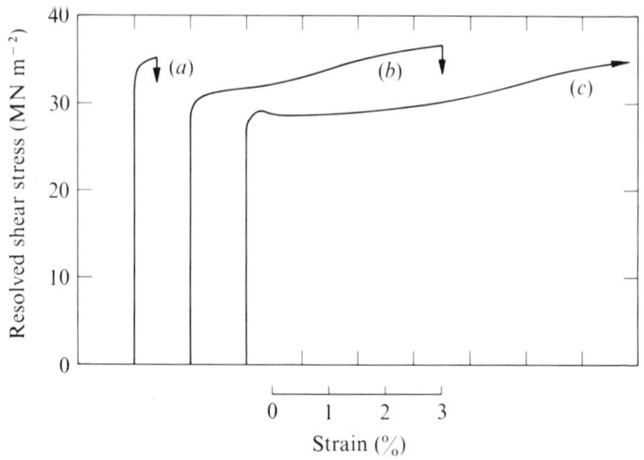

Fig. 4.4. Tensile stress/strain curves for MgO single crystals at room temperature: (*a*) few intersecting slip bands; (*b*) many intersecting slip bands, (*c*) single expanding slip band. (After Stokes *et al.*, 1961.)

The stress to nucleate the crack can be estimated approximately following an argument of Petch (1968). According to the calculations of Eshelby *et al.* (1951), a tensile stress $N\tau$ exists near the head of a pile-up of N dislocations under a shear stress τ. The stress also has to overcome a lattice resistance stress τ_0 and thus the total stress compressing the dislocations is $N(\tau-\tau_0)$. Fracture should initiate when this stress reaches the theoretical strength of the material and this leads to the critical crack-initiation condition

$$N(\tau-\tau_0)=\sigma_{th} \tag{4.1}$$

and it remains to eliminate N. The elastic shear displacement across the slip band of length l_s is $(\tau-\tau_0)\,l_s/G$ and this is equivalent to a plastic displacement Nb. Thus

$$Nb=(\tau-\tau_0)l_s/G. \tag{4.2}$$

Substitution into (4.1) thus gives the critical condition

$$(\tau-\tau_0)^2=Gb\sigma_{th}/l_s. \tag{4.3}$$

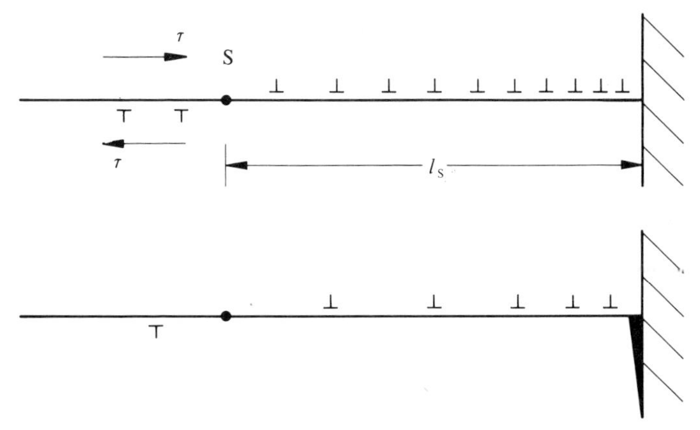

Fig. 4.5. Zener (1948) model for nucleation of a crack by the pile-up of dislocations at a barrier.

This can be arranged in a more useful form by making a number of substitutions. For the grains most favourably oriented for slip (i.e. at 45° to the applied stress) the shear stress is half the applied stress, $\tau = \sigma/2$; $E = 2G(1+\nu)$, (2.2). Finally, recalling (3.6) and (3.7), the approximations $\sigma_{th} \sim (E\gamma_0/b)^{\frac{1}{2}}$, and $\gamma_0 \sim Eb/125$, letting $r_0 = b$, (4.3) can be rearranged as

$$\sigma \sim \sigma_0 + \left[\frac{20E\gamma_0}{(1+\nu)l_S}\right]^{\frac{1}{2}} = \sigma_0 + k^* l_S^{-\frac{1}{2}}. \tag{4.4}$$

This shows that the crack-initiation stress is dependent upon a term for the length of the slip band; this can be related to various microstructural features such as the spacing between slip bands or the grain size. Note that the proportionality factor k^* has the same dimensions as the fracture-mechanics stress-intensity factor K; k^* can thus be regarded as a microscopic stress-intensity factor.

We can now return to the varied behaviour of single crystals as depicted in fig. 4.4. Numerous slit cracks, believed to have been formed by the mechanism described above, have been observed in deformed magnesia single crystals. Figure 4.6 sketches the relative position of such flaws in a crystal that has deformed on two orthogonal slip planes. Here we see that the barrier to slip is represented by the more strongly established slip bands and that the cracks form along the boundary of these bands. It should be noted particularly that the length of the crack tends to be limited by the spacing between the narrower slip bands. These cracks are often just a few μm long and stable, being too small to propagate under stresses near the flow stress of the material, as discussed earlier for the small, inherent cleavage cracks. The cracks that form in crystals with a high density of slip sources are thus limited in this way and such crystals show a moderate ductility. In contrast, crystals slipping on relatively few slip planes do not enjoy the crack-length-limiting effect of nearby parallel slip planes and thus the crack grows to a sufficient size to propagate catastrophically after relatively small plastic strains. At the other extreme crystals slipping on just one slip band have no crack nucleation mechanism and are thus very ductile.

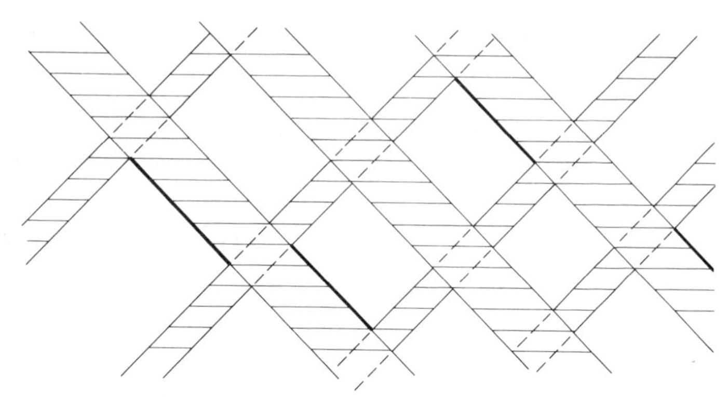

Fig. 4.6. Location of {110} cracks along the edges of wide slip bands, believed to have been nucleated by pile-ups of dislocations on the narrower orthogonal slip bands. (After observations of Stokes *et al.*, 1959.)

The important observation from these experiments is that even in single crystals the stress concentrations set up due to plastic deformation can readily lead to crack initiation and premature failure. This is associated with the relatively few available slip systems in ceramics; in metals the stress concentrations are usually relieved by slip on more numerous and alternative planes. The slip systems in ceramics have an even more restrictive effect in polycrystals and this is discussed below.

4.3 Plasticity of polycrystals

Although single crystals of some ceramics such as MgO show ductility at low temperatures for specific conditions of slip geometry, these materials in polycrystalline form are invariably brittle under tensile stresses (plastic strain at failure $\leqslant 0.1\%$). This is a direct consequence of the number of slip systems available.

In order for a polycrystalline material to deform plastically in response to a specific applied stress, then each individual grain must be capable of undergoing an arbitrary change in shape, otherwise voids or cracks forming at the grain boundaries would eventually lead to failure. Slip occurs first in those grains where the resolved shear stress on the slip planes is a maximum. The neighbouring grains are in different orientations and thus attempt to deform by slip on systems in different orientations.

Von Mises (1928) first showed that a general homogeneous strain by slip required the operation of five independent slip systems. A small cubic element of material under an arbitrary state of strain requires definition of six strain components (§ 2.1). These are the tensile (or compressive) strains in three orthogonal directions plus the shear strains on three orthogonal planes, fig. 2.1. Because during plastic deformation there is usually no change in volume the sum of the three tensile (or compressive) strains is zero. This reduces the six independent components of the general strain to five. Von Mises noted that the operation of one slip system produces just one independent component of strain. Thus five independent slip systems are needed to produce a general homogeneous strain by normal plastic deformation. A slip system is independent provided that its operation produces a change in shape which cannot be produced by a combination of slip on the other independent systems.

Determination of the number of independent slip systems for a particular set of slip systems in a crystalline lattice is a straightforward but detailed process best considered by a strain tensor analysis. Groves and Kelly (1963) describe this for a number of common crystal systems. Face-centred-cubic metals, for example copper, slip on the $\{111\}\langle 1\bar{1}0\rangle$ family of slip systems and have five independent slip systems out of the possible twelve. On the other hand, magnesia, which slips on the slip systems $\{110\}\langle 1\bar{1}0\rangle$, has only two independent slip systems out of a possible six. Data for several ceramics are included in table 4.2. This shows that at low temperatures either dislocations are not mobile (to be discussed in chapter 5) or the number of independent systems is less than five, so that general strain changes cannot be accommodated by plastic deformation. The stress concentrations that build up at the ends of slip bands, in those grains most favourably oriented for slip, can thus nucleate cracks by the mechanism illustrated in fig. 4.5 where the slip barrier now takes the form of a grain boundary.

Similar arguments to the above apply to the blunting of cracks. To blunt a crack of arbitrary shape by inducing an increase in the radius of curvature at the crack tip requires an arbitrary change of shape by slip. It follows, therefore, that ductile metals like copper also have the capability of blunting cracks; any cracks are thus easily rendered harmless and extensive plastic flow can then occur. On the other hand, cracks in brittle ceramics generally cannot be blunted by plastic flow and thus the stress-concentrating effect of the cracks leads to brittle failure.

At high temperatures additional, or secondary, slip systems can operate and for magnesia these are $\{001\}\langle 1\bar{1}0\rangle$. Combined with the primary slip systems there are now a total of five independent slip systems. (Note that it is not always possible to simply sum the numbers of independent slip systems for the two sets of systems to obtain the total number of

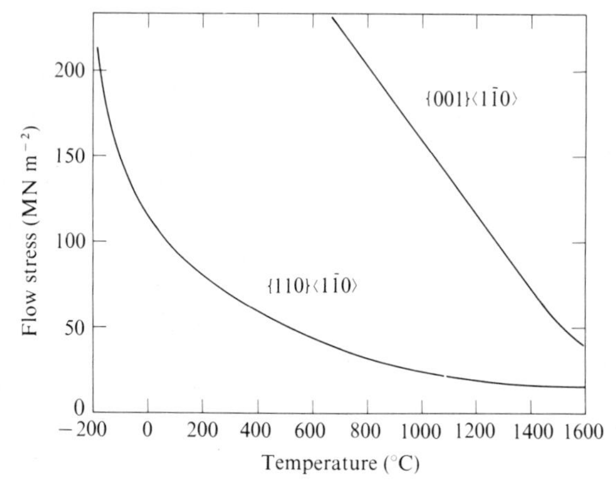

Fig. 4.7. Temperature-dependence of the flow stress of single crystals of MgO, deformed in $\langle 100\rangle$ for $\{110\}\langle 1\bar{1}0\rangle$ slip, and $\langle 111\rangle$ for $\{001\}\langle 1\bar{1}0\rangle$ slip. (After Hulse and Pask, 1960.)

independent systems; the total may be less if the two sets of systems are related in specific geometrical ways.) Magnesia is thus ductile at temperatures greater than 1700 °C. The same transition occurs in polycrystalline NaCl but at ~200 °C.

The ductile-to-brittle transition temperatures do not correspond to the sudden possiblity of slip on the secondary system. Slip on both the primary and secondary systems is a highly temperature-dependent process, as we shall discuss in chapter 5. Figure 4.7 shows the temperature dependence of the flow stress of single crystals of magnesia for slip on the $\{110\}\langle 1\bar{1}0\rangle$ and $\{001\}\langle 1\bar{1}0\rangle$ systems. The former were obtained by deforming crystals along a $\langle 100\rangle$ direction and the latter by deforming along a $\langle 111\rangle$ direction. In the latter case there is no resultant shear stress on the primary slip system and thus secondary slip can occur, albeit at high stresses, at temperatures as low as 800 °C. The important point, however, is that in a polycrystal, slip on the primary system results in stresses at grain boundaries at quite low stresses and this leads to grain-boundary cracking before relief can be obtained by slip on the secondary system.

Day and Stokes (1966) conducted a detailed series of tensile tests on polycrystalline magnesia specimens from a number of commercial sources and also on specimens produced by a recrystallisation treatment. The latter specimens were prepared by extending crystals with a ⟨110⟩ tensile axis at 1800 °C by ~60%. They had noted previously that such a treatment produced material which had a very high dislocation density and which could be made to recrystallise at ~2000 °C to produce a coarse grain size and uniform polycrystalline structure. Further tensile tests on the recrystallised specimen showed that at 1600 °C and below, the plastic deformation was very small and certainly <1% before brittle fracture. At 1700 °C, ~10% plastic elongation preceded brittle fracture but at 1800 °C and above the specimens were completely ductile and necked down to a point after plastic strains as high as 100%. Interestingly, Day and Stokes observed that a brittle-to-ductile transition could be obtained only in some commercial hot-pressed materials and then at temperatures ~2100 °C or above. This difference was attributed to both the porosity and impurities situated at the grain boundaries in the hot-pressed materials, in contrast with the boundaries in the recrystallised material that were free from pores and impurities. The defects in the grain boundaries in the hot-pressed materials were considered to promote grain-boundary sliding which led to premature failure (see § 5.3).

In conclusion we see that even in those ceramics which show some plasticity in single-crystal form, the geometrical constraints restrict slip in polycrystals. Very few ceramics are significantly plastic even at high temperatures although MgO and UO_2 are noticeable exceptions. In some crystals with cubic symmetry (for example spinel and β-SiC) that slip on {111}⟨1$\bar{1}$0⟩ systems the geometrical constraint is absent and five independent systems are available; unfortunately, as we discuss later, the covalent bonding in these materials means that the dislocations are very immobile.

5 The flow stress and plastic flow

In the previous chapter, arguments were developed to show how geometrical aspects of plastic deformation by dislocation motion could explain the general lack of ductility in ceramics and hence the usual brittle behaviour. Little comment was made on the mechanisms that control dislocation flow stress in ceramics but the important conclusion was that, even when dislocation motion was possible, restrictions on the number of active slip systems reduced the ductility of polycrystals to an extremely small level. In this chapter we discuss in outline the factors that control the dislocation flow stress and consider which other mechanisms of plastic flow are important.

Plastic flow in crystalline solids involves the transport of material to produce a change of shape. This may be achieved by dislocation motion, by the movement of individual atoms through diffusional processes or by shear on grain boundaries. Plastic flow is influenced by several factors as summarised in table 5.1.

The first factor, the lattice structure and type of bonding, is the most important in that it establishes the base behaviour of the ceramic in question. The defects, or departures from the perfect lattice, have been divided into two categories: intrinsic defects which occur in pure materials; and extrinsic defects, or impurities, which occur in all practical materials. With so many influencing factors, plastic flow in ceramics is expected to exhibit a rich variety of behaviour; here, we present only the broad principles.

In chapter 4 we showed why even in ceramic crystals where plastic flow was possible, the amount of flow was generally small because this led to the generation of cracks and premature failure. A key parameter is thus the flow stress. To suppress the latent possibility of brittle fracture it is usual (and simpler) to investigate the plastic-flow behaviour of ceramics by means of compression tests. The flow stress may be quoted simply as the applied stress, or, when dislocation motion is dominant, as the critical resolved shear stress – that is the applied stress resolved on the active slip plane in the active slip direction.

We examine first the case where dislocation behaviour is the controlling factor. Figure 5.1 shows how the critical resolved shear stress varies as a function of temperature, where the shear stress has been normalised by dividing by the shear modulus and the temperature quoted in terms of the fraction of the melting temperature. The deformation effects often depend upon strain rate and behaviour for two strain rates is indicated. The scheme may be divided into three main regions, I, II and III with decreasing values of τ/G.

Table 5.1. *Factors affecting the flow stress and plastic flow of ceramics*

Factor	Type of defect	Examples
Lattice structure and type of bonding	–	–
Intrinsic lattice defects	Point defects, larger defects, dislocations, grain boundaries	Vacancies, interstitials, voids, pores
Extrinsic lattice defects	Impurities	In solution, at grain boundaries, discrete second phase

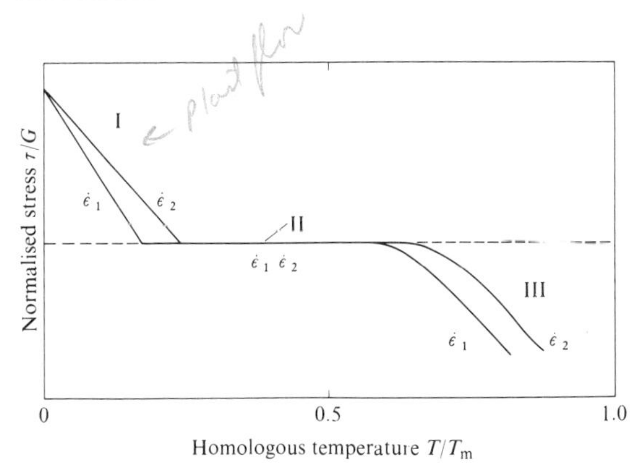

Fig. 5.1. Schematic variation of the critical resolved shear stress with temperature at two strain rates, $\dot{\varepsilon}_2 > \dot{\varepsilon}_1$. (After Evans and Langdon, 1976.)

In region I, τ/G decreases rapidly with increasing temperature and, for a given temperature, is greater the higher the strain rate. Plastic flow in region I is thermally activated and is controlled by thermal fluctuations which allow dislocations to overcome the short-range stress fields of obstacles in the glide plane. The obstacles may be associated with the lattice structure itself or with particular lattice defects. In region II, τ/G is constant and independent of both temperature and strain rate. Here behaviour is athermal and flow is controlled by the resistance arising from the long-range stress fields of particular defects, such as impurity precipitates, or parallel dislocations on nearby slip planes.

In region III, τ/G again falls with increasing temperature and a similar strain-rate dependence to that in region I is observed. At the high temperatures involved, various creep processes not necessarily entailing dislocations are now possible and this produces a flexibility in plastic behaviour that may relax the purely geometrical restriction described in chapter 4. More appreciable plastic flow can thus take place without necessarily leading to the formation of crack-like defects and premature brittle failure.

Our aims here are first to explain why dislocation motion is relatively easy in some materials, for example MgO, more difficult in others, for example Al_2O_3 and very difficult in others, such as SiC and Si_3N_4; secondly, to discuss the influence that lattice defects have in controlling the flow stress at lower temperatures; and finally to consider the main creep-controlling mechanisms to see how behaviour is affected when second-phases are present.

5.1 Dislocation motion in pure single crystals

The stress to move glide dislocations through a crystalline lattice at a particular temperature is dependent on a number of factors including the elastic properties of the material, the type of bonding and the detailed structure of the core of the dislocations. In discussions of the slip systems in crystals we saw that the atoms along the line of the dislocation are situated successively at sites of relatively low and high distortion as the dislocation moves through one Burgers vector. The energy associated with the dislocation thus varies with distance depending on the arrangement of the atoms near the core of the dislocation in the mid-position. An applied stress, usually refered to as the Peierls–Nabarro stress, is required to overcome this periodic variation in energy. For dislocation glide in MgO on the $\{110\}\langle 1\bar{1}0\rangle$ system the distortion at the mid-position is much less than for $\{001\}\langle 1\bar{1}0\rangle$ glide; this is related to the respective atomic positions near the cores of the dislocations for the two systems indicated in fig. 4.1. The energy of these two types of dislocation varies with displacement in a different way, and this is shown schematically in fig. 5.2. The Peierls–Nabarro stress is thus low for $\{110\}\langle 1\bar{1}0\rangle$ glide in MgO and relatively high for $\{001\}\langle 1\bar{1}0\rangle$ glide.

Apart from variations in Peierls–Nabarro stress for different dislocations in a particular crystalline lattice, this stress also varies significantly with the type of bonding. The key parameter here is the width of the dislocation which is defined as the width of the region in the slip plane, and measured perpendicular to the dislocation line, within which the displacement of the atoms on either side of the slip plane exceeds one half the maximum value. At one extreme the atoms in metals are bonded primarily by a free-electron gas and thus, provided that the interatomic spacing is preserved, the bond energy is relatively insensitive to the angle between the atoms. At the other extreme, the strongly directional bonds in covalent materials are responsible for large increases in energy when

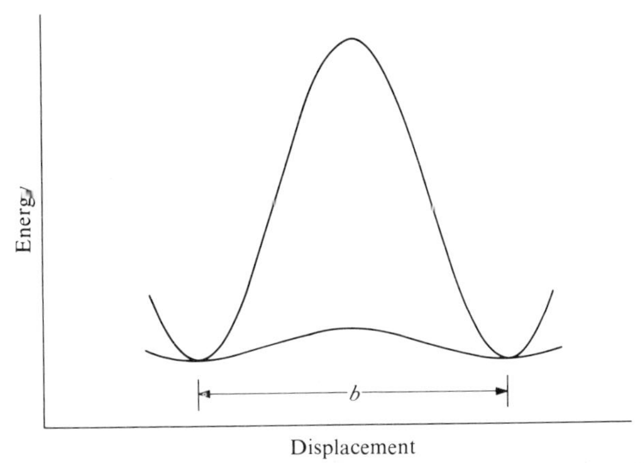

Fig. 5.2. Variations of dislocation energies with dislocation position for dislocations of low and high Peierls–Nabarro stress.

the angle of the bond varies. Dislocations in metals are thus relatively wide and in covalent crystals very narrow. With a wide dislocation the total distortion is spread over a relatively large volume and any variations in energy with dislocation position are small; this results in a low Peierls–Nabarro stress. In ductile metals the width of dislocation is of the order of ten atomic spacings, whereas in covalent crystals it is only one or two atomic spacings. The bonds in ionic crystals, which are also non-directional, again produce a tendency for wide dislocations and a low Peierls–Nabarro stress.

We saw in chapter 1 that bonding in ceramics may be primarily ionic as in MgO or primarily covalent as in SiC, although pure types of bonding are rare and most materials show hybrid behaviour. Precise calculations of the Peierls–Nabarro stress are particularly difficult in that they involve estimations of the energy associated with the distortions near the core of the dislocation; these distortions are much too great to be considered by linear-elasticity methods. Simplified calculations have been performed by Sanders (1962) and discussed by Kelly (1973) and refer to a primitive cubic crystal. Their argument shows in a qualitative way how the Peierls–Nabarro stress varies with type of bonding. The calculation indicates that the Peierls–Nabarro stress τ_{PN} is related to the dislocation width w by

$$\tau_{PN} \propto \exp(-w). \tag{5.1}$$

Thus τ_{PN} is highly dependent on w, in accordance with the observation that dislocations in MgO, which has ionic bonding and wide dislocations, are mobile at ambient temperatures and above but those in SiC, which has covalent bonding and narrow dislocations, are mobile only at temperatures >2000 °C.

Unfortunately most calculations of the Peierls–Nabarro stress for ceramic-type materials are limited to the alkali halides where there is a sound model for the ionic bonding forces. The factors controlling the flow stress of the more refractory ceramics are not well understood although it has been suggested by Conrad (1965) that the flow stress of single-crystal sapphire at temperatures <1500 °C is controlled by the Peierls–Nabarro stress.

5.2 Effects of lattice defects on the flow stress of single crystals

The Peierls–Nabarro stress, being dependent on the type of crystalline lattice, determines only the minimum force to move dislocations. Other factors are of considerable additional importance, especially in crystals with ionic bonding and a small Peierls–Nabarro stress. In pure crystals these include interactions with point defects, other dislocations and grain boundaries, and, in impure crystals, with the impurities which may be distributed throughout the lattice or be present as a separate phase.

The effects of impurities on the flow stress are particularly diverse. Table 5.2 gives data for MgO containing just 0.5% iron impurity by weight; the flow stress varies by an order of magnitude depending on the form of the iron impurity. When the iron is in solution with the same valency as the magnesium ions the flow stress is low. When the iron is oxidised to Fe^{3+} one charge-compensating cation vacancy is required for each two iron ions in solution; both the impurity ions and the vacancies introduce a charge imbalance into the structure and this is associated with an increase in flow stress. The highest flow stress is obtained when a fine dispersion of spinel precipitates is produced.

All crystals contain impurities and those of commercial origin typically from 0.1 to 1%. The effects indicated in table 5.2 are thus common to a wide range of materials. When the impurities in ionic crystals are in solid solution and have the same valency as the host crystal, the hardening effects are generally small. Conversely, impurities of different valency show pronounced hardening effects. For a given concentration of impurity the hardening in the latter case is usually greater by one-to-two orders of magnitude. The amount of hardening produced when the impurities are present as a discrete second phase depends on the size and concentration of impurity. A high concentration of small impurity precipitates produces the maximum hardening.

Table 5.2. *Variation of the flow stress of single-crystal MgO containing 0.5% iron by weight at 20 °C (Davidge, 1967)*

Treatment	Form of iron	Critical resolved shear stress ($MN\ m^{-2}$)
1400 °C in CO + rapid cool	Single Fe^{2+} ions	30
1400 °C in air + rapid cool	Single Fe^{3+} ions + vacancies	130
1400 °C in air, rapid cool + 1 h at 700 °C	$MgO.Fe_2O_3$ precipitates	300

There are two main theories describing the hardening due to impurities in solid solution based on elastic or electrostatic models. The elastic theory (Fleischer, 1962) has been applied successfully to simple systems, for example magnesium impurity in lithium fluoride. Here the magnesium ions are present as dipoles with a charge-compensating vacancy in an adjacent cation site. The essence of the theory is that the dipoles introduce strong and asymmetrical elastic distortions into the lattice which impede the motion of dislocations passing within an atom spacing of the defect. The force to move the dislocation past the defect is supplied partly by an applied stress and partly by thermal activation. The force/distance relations proposed rise very steeply as the defect is approached by the dislocation and there is a strong temperature dependence of strength. Experiment and theory are in reasonable agreement, with the flow-stress contribution τ^* due to this hardening varying with temperature as $\tau^{*\frac{1}{2}} \propto (1 - T^{\frac{1}{2}})$, and with impurity concentration C^* as $\tau^{*\frac{1}{2}} \propto C^{*\frac{1}{2}}$. The electrostatic theory (Gilman, 1974) produces quantitatively similar conclusions but assumes that the force to shear through a dipole is similar to the adhesive strength of the defect. An attraction of this approach is that it explains successfully that hardening is associated with the charge difference between the impurity and host crystal and not the properties of the impurity and host ions. These theories are not yet applicable to the situation where the simple defects begin to agglomerate into larger defects, before discrete precipitates appear.

When second-phase precipitates are present, the hardening is greater (except at very low temperatures) than when the impurities are in solid solution, see fig. 5.3 for LiF containing Mg. At low temperatures and high stresses, in region I (fig. 5.1), the force/distance curve as the dislocation approaches a precipitate is less steep and the strain field round the defect less localised than for a single impurity ion; the temperature dependence of the strength is thus less in the former case.

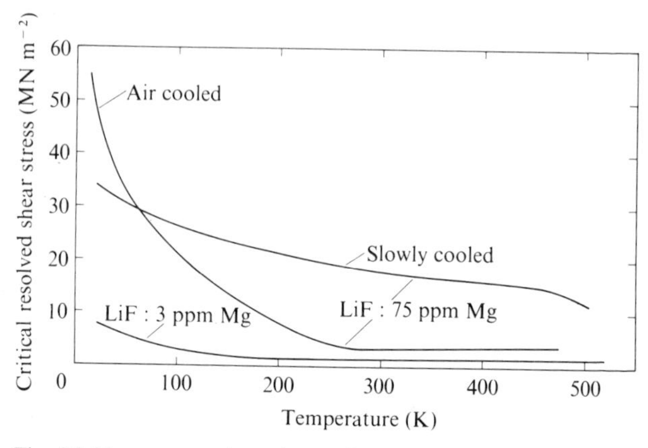

Fig. 5.3. Temperature dependence of the critical resolved shear stress for LiF crystals of two purities. Slow cooling leads to the formation of precipitates whereas with air cooling impurities are retained in solution. Data for the purer crystal do not depend on cooling rate. (After Johnston, 1962.)

Region II in figs. 5.1 and 5.3 occurs at higher temperatures and the force required for the dislocations to cut through the precipitates is higher than for single ions. In many respects the precipitation hardening in ceramics displays the classical behaviour shown by metals. As the precipitate size grows the strength first increases and then falls again once the particles are larger than a critical size. The quantitative aspects of this hardening, however, have only been considered for very few ceramic systems.

The above effects are only of major practical significance in materials with ionic bonding and a low Peierls–Nabarro stress. Similar effects occur in solids with more covalent bonding, but generally at temperatures >1000 °C. At these temperatures, however, the effects of grain boundaries become particularly important with reference to plastic deformation and we consider next polycrystals with special emphasis on creep processes.

5.3 Creep and plastic deformation in polycrystals

At high temperatures (region *B* in fig. 1.13) limited plasticity is possible in ceramics and this may have several origins including dislocation motion, grain-boundary sliding or softening of minor phases. Generally, significant plasticity is not observed and the almost universal consequence is that cracks are eventually nucleated at the boundaries.

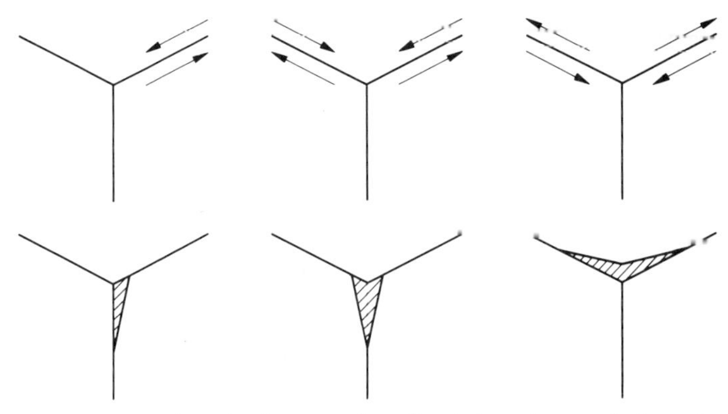

Fig. 5.4. Mechanisms for the generation of grain-boundary cracks by grain-boundary sliding. (After Vasilos and Passmore, 1968.)

Some of the relevant properties of grain boundaries were mentioned in § 1.4. Boundaries can act as sources and sinks for point defects and often attract impurities or are the sites of second phases. At temperatures where point defects are mobile plastic flow should, therefore, occur preferentially at grain boundaries. This can lead to several phenomena such as changes in the shape of individual grains or the sliding of adjacent grains relative to each other across the boundary. There are numerous ways in which grain boundary cracks may be nucleated by grain-boundary sliding and some of these are illustrated in fig. 5.4. An observation is shown in fig. 5.5 where cracks have developed in an MgO polycrystal deformed under a tensile stress at 1200 °C. Clearly the presence of such cracks, which often link up along adjacent grain boundaries, can lead to premature brittle fracture. When second phases are present at the boundaries the dihedral angle (see § 1.4) is very important. When this is low the second phases have high penetration along the boundaries, as in fig. 1.10. The second phases have a much greater effect in materials of this type of microstructure at high temperature than when the dihedral angle is large.

The plastic processes occurring prior to failure at high temperatures are referred to as creep. Figure 5.6 indicates the typical creep behaviour

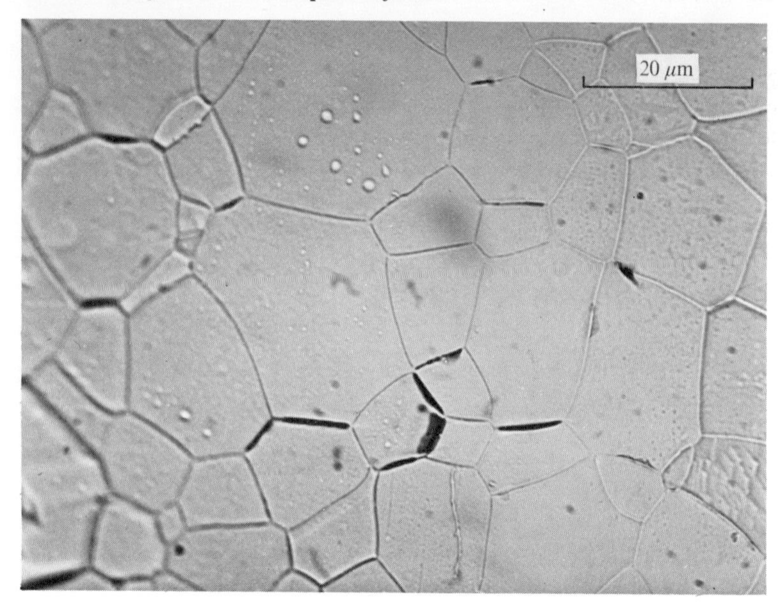

Fig. 5.5. Grain-boundary cracks in MgO formed under a tensile stress at 1200 °C. The stress direction is vertical; note that the cracks form predominantly normal to this direction. (After Evans *et al.*, 1970.)

for ceramics expressed in terms of strain v. time under a constant stress. The curve is in three sections. After an instantaneous deformation on application of the stress, the creep rate decreases with time during the first stage and then reaches a steady state in region II. Under a tensile stress brittle failure often intervenes in this stage, although under compressive stress a third stage is usually observed. In tension the total strain is usually $<1\%$.

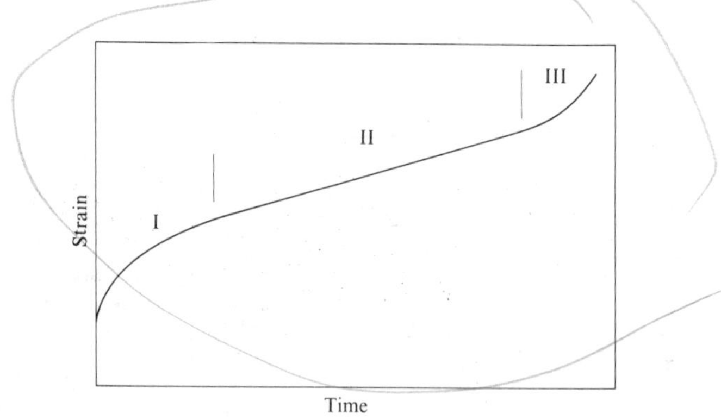

Fig. 5.6. Typical creep curve for ceramics.

In polycrystals the underlying creep mechanisms are of two types: those concerning the grain boundaries and those within grains. Fortunately almost all mechanisms obey a general relationship for steady-state creep of the form

$$\dot{\varepsilon}=\frac{A^{*}DGb}{kT}\left(\frac{b}{l_g}\right)^{m^*}\left(\frac{\sigma}{G}\right)^{n^*}. \tag{5.2}$$

A^* is a dimensionless constant, D a diffusion coefficient (having an exponential temperature dependence), G the shear modulus, b the magnitude of the Burgers vector, l_g the grain size, σ the applied stress, and m^* and n^* constant exponents. There is a large number of mechanisms and

Table 5.3. *Creep equation constants and diffusion paths for various creep mechanisms. (After Langdon, 1975.)*

Creep mechanisms	m^*	n^*	Diffusion path
Lattice mechanisms			
Dislocation glide/climb, climb controlled	0	4.5	Lattice
Dislocation glide/climb, glide controlled	0	3	Lattice
Dissolution of dislocation loops	0	4	Lattice
Dislocation climb without glide	0	3	Lattice
Dislocation climb by pipe diffusion	0	5	Dislocation core
Boundary mechanisms			
Vacancy flow through grains (Nabarro–Herring creep)	2	1	Lattice
Vacancy flow along boundaries (Coble creep)	3	1	Grain boundary
Grain-boundary sliding with liquid phase	1	1	Second phase
Grain-boundary sliding without liquid phase	1	2	Lattice or grain boundary

predicted values for the exponents m^* and n^* are summarised in table 5.3. In principle, therefore, it is possible to identify the creep mechanism from analysis of basic creep data. In practice this may be difficult if two or more processes occur simultaneously or sequentially, and because of problems in obtaining reliable values for A^* and D in (5.2). As a simple example, an individual grain in a polycrystal could change its shape by a vacancy diffusion mechanism either through diffusion along the boundary (Coble creep) or through the grain (Nabarro–Herring creep); these processes may be distinguished experimentally by investigating the grain-size dependence of creep and calculating the exponent m^*.

For a small number of ceramics (mainly pure oxides) understanding is sufficiently advanced to present detailed information about the dominant creep mechanisms in particular regimes of stress, temperature and grain size. This data is conveniently expressed in the form of a *deformation mechanism map* (Ashby, 1972). Essentially this involves generation of constitutive equations in the form of (5.2) for the relevant creep mechanisms, which are then used to plot the map under conditions of either constant grain size or constant temperature. An example of the latter type is given in fig. 5.7 for $MgO.Al_2O_3$ spinel at 1500 °C. Constant-strain-rate contours are superimposed so that it is possible to determine the creep rate as well as the creep mechanism for specific combinations of stress and grain size.

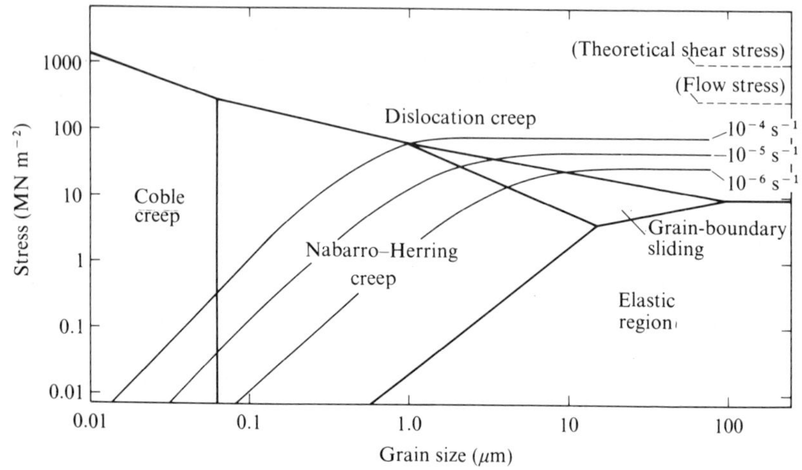

Fig. 5.7. Deformation map for $MgO.Al_2O_3$ spinel at 1500 °C. (After Notis, 1975.)

The theoretical importance of deformation mechanism maps is thus obvious but their potential, particularly for commercial materials, is dependent on major advances in generation of experimental data and in fundamental understanding. For example, only limited creep data is available for the extremely important class of refractory materials which have both complex microstructures and phase relationships. Another area where further theoretical developments are required is the prediction of fracture behaviour from creep data. Many of the creep processes described above lead to grain-boundary cracks or voids but how this relates to the ultimate failure condition is not clear.

The second phases in ceramics are usually less refractory than the host material and their plastic properties are thus likely to dominate the high-temperature behaviour. The intrinsic properties of dislocations in the host lattice may be of minor significance, particularly in covalent crystals. Examples for specific materials to be discussed in the next chapter illustrate this.

6 The fracture strength of ceramics

We can now bring together the threads of the arguments developed in previous chapters and explain in a quantitative way the strength of ceramics in terms of fundamental parameters such as the inherent flaw size, the effective surface energy and the flow stress of the material. All practical ceramics are brittle, except at high temperatures, but an important distinction in behaviour can be made between fracture originating at the inherent flaws in the material and fracture originating from flaws generated by limited-plastic-flow processes. Even when plastic flow is possible the amount is quite small and we will show that the flaws generated by the plastic deformation can often grow to such a size that the Griffith equation is satisfied at stresses close to the flow stress.

Figure 6.1, which is an elaboration of fig. 1.1, indicates some of the more important textural features which are controlled by the fabrication conditions and that affect both the flow stress and fracture stress of the material. External factors of major significance include the temperature and the environment. The general form of the temperature dependence of strength of ceramics was indicated in fig. 1.13 and it will become clear that the relatively temperature-independent region of strength at low

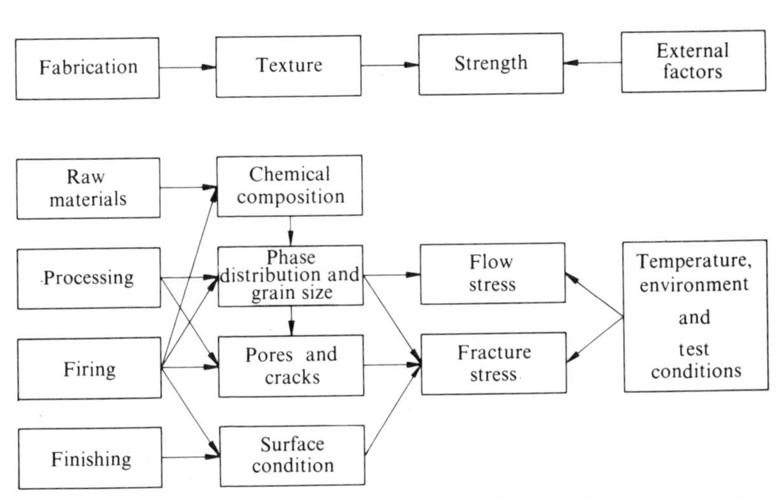

Fig. 6.1. The major fabrication and textural factors affecting the strength of ceramics. (After Davidge, 1972.)

temperatures is controlled by the propagation of fracture from inherent flaws, whilst the temperature-dependent part of the curve at higher temperatures is controlled by fracture initiated at cracks produced by limited plastic flow. Strength can thus be understood in terms of (3.17). We discuss here the surface-energy values typically found for ceramics, the origins of flaws, and then present strength data for a selection of engineering ceramics.

6.1 Surface-energy data

Single crystals

The energy associated with the surface of a material arises from differences between the forces acting on the atoms at the surface and those on the atoms in the bulk of the material. Energy is required to create new surface and this may be supplied in several ways: mechanical forces, chemical solution, or vaporisation at high temperature. In each case, energy is absorbed through the breaking of the bonds between the atoms. Initially, we give attention to the surface energy of ceramics in single-crystal form both from a theoretical and experimental standpoint. We are concerned with the surface free energy γ_0 defined as the reversible free-energy change during the formation of unit surface area under defined conditions including temperature, pressure and orientation. There are several ways of estimating surface-energy values from thermodynamic data, and methods have been discussed by Bruce (1965) and Nurse (1968). Bruce, for example, uses a simple method involving the sublimation energy and an approximate relationship

$$\gamma_0 = \gamma_0^* - \Gamma T, \tag{6.1}$$

where γ_0^* is the surface energy at 0 K, Γ is a constant and T is the absolute temperature. Bruce's values for a number of common ceramics are included in table 6.1. The surface free energy is a minimum for a particular crystallographic plane, as indicated, and both the minimum and the average values for temperatures of 20 °C and 1500 °C are tabulated.

The surface-free-energy values are typically 0.5–3 J m^{-2} and there are modest variations with both temperature and crystal orientation. The important point, however, is that these values relate to ideal reversible cleavage processes which are rarely observed under practical conditions. Deviations in both directions from the ideal are observed. In many cases other energy absorbing mechanisms operate and experimentally measured values are greater than calculated values. At the other extreme, low values can be recorded when, for example, single crystals are tested under chemically corrosive conditions, including water for oxides, so that the atomic bonds are broken with the assistance of chemical forces.

An extensive series of experimental measurements of surface energy for single crystals has been made by Gilman (1960) using the double-cantilever beam specimen described in chapter 3. The values are thus for an effective surface energy γ_i to initiate cleavage fracture, which may or may not correspond to the thermodynamic data. Gilman obtained a

Table 6.1. *Surface-free-energy values (mJ m^{-2}) calculated from sublimation-energy data. (After Bruce, 1965)*

Material	Minimum energy plane	Minimum surface free energy (γ_m)	Average surface free energy (γ_a)	γ_m		γ_a	
				20 °C	1500 °C	20 °C	1500 °C
MgO	{100}	$1680-0.308T$	$2600-0.476T$	1590	1150	2460	1756
BeO	{0001}	$2140-0.321T$	$2400-0.359T$	2050	1570	2290	1760
β-SiC	{110}	$2030-0.370T$	$3000-0.546T$	1920	1370	2840	2030
SiO_2	{110}	$616-0.128T$	$925-0.193T$	580	390	870	580
TiO_2	{110}	$426-0.089T$	$800-0.167T$	400	270	750	500
Al_2O_3	$\{10\bar{1}4\}$	$1120-0.216T$	$1200-0.232T$	1060	740	1130	790
Cr_2O_3	$\{10\bar{1}4\}$	$857-0.186T$	$925-0.200T$	800	530	870	570

selection of single crystals that exhibited well-defined cleavage planes and arranged the specimen geometry so that the cleavage crack propagated along the centre of the cantilever beam. Crystals included LiF, MgO, CaF_2, BaF_2, $CaCO_3$, Si and Zn. Sharp cracks were introduced into the specimens, with a wedge device similar to that described in chapter 3, in liquid nitrogen to suppress plastic deformation. Data for magnesia illustrate the typical effects observed. When the specimen was precracked at -196 °C and immediately tested at this temperature, mean values of effective surface energy of ~ 1.2 J m^{-2} were obtained, closely in line with thermodynamic data. However, other samples that were precracked and/or tested at 25 °C, gave γ_i values significantly higher, up to 3 J m^{-2}. This was attributed to a contribution to the effective surface energy of plastic deformation near the crack tip, which was revealed by dislocation-etching techniques. The important conclusions of this work are that in careful experiments where fracture is carried out in an almost ideal reversible manner the experimental data are in accordance with expectations from thermodynamic theory. However, it is vital to suppress plastic deformation otherwise much higher values of γ_i are observed.

Similar measurements have been carried out by Wiederhorn (1969) on alumina, a material that does not exhibit a well-defined cleavage plane. Successful measurements were possible only for certain crystal planes and values of 7.3 and 6.0 J m^{-2} were obtained for the $\{10\bar{1}0\}$ and $\{\bar{1}012\}$ planes. Data could not be obtained for the $\{0001\}$ planes because of their lack of charge neutrality. These values are considerably higher than the thermodynamic estimates for the surface energy given in table 6.1, differences being attributed to possible non-conservative effects during fracture, such as heat generation or plastic deformation near the crack tip, although no positive evidence for the latter could be obtained experimentally. Additional measurements were made by Wiederhorn *et al.* (1973) as a function of temperature. Careful transmission-electron-microscopy studies of the regions near the tips of arrested cracks showed no evidence for plastic deformation processes at temperatures < 400 °C, although evidence was obtained at higher temperatures but only in a small proportion of the specimens examined. For some single crystals there is thus a lack of precise understanding of the energy-absorbing processes during cleavage fracture.

Polycrystals

There is now extensive information available on experimental values for γ_i for a wide range of polycrystalline ceramics. Data for a number of specific materials will be discussed later in this chapter, but values of 10–50 J m^{-2} are commonly observed. These are about an order of magnitude greater than values for single crystals. The fracture process for a polycrystal must necessarily be more complicated than that for a single crystal. The most obvious effect is that the fracture surface of a polycrystal is much rougher than that of a single crystal. This means that the effective surface energy for polycrystals is much higher because fracture energy values are expressed per unit planar projected area. The rough surface is associated with the necessary changes in orientation of the crack as it passes from grain to grain or from grain boundary to grain boundary. A second reason for the relatively high γ_i values for polycrystals is that the presence of plastic deformation in zones adjacent to the fracture face is much more common than for single crystals; this is again associated with the increased complexity of the fracture process. Other factors include subsidiary cracking which may be connected to the main fracture crack and various less-easily quantified effects, including the generation of energy through heat, light or sound as the crack propagates.

Evans (1970) working with large-grain-size MgO polycrystals performed a careful assessment of the various factors contributing to the experimental effective-surface-energy value according to

$$\gamma_i = \eta\gamma_0 + \gamma_p + \gamma_u, \qquad (6.2)$$

where η represents a factor due to the non-planar surface, γ_p the contribution from plastic deformation, and γ_u an unquantifiable factor. Evans found that the bulk of the surface-energy value could be accounted for by plastic deformation near to the fracture face with a lesser contribution due to the roughness of the surface. γ_u represented a small fraction of the total surface energy and thus as in single-crystal MgO there is good fundamental understanding. For reasons not fully clear at present γ_u appears to be small for ionically bonded crystals and large for covalent crystals.

It follows from this argument that it is very difficult, if not impossible, to estimate a surface-energy value for a material prior to fracture, but only to perform a post-mortem analysis as described above. Nevertheless it might be expected that γ_i varies in a systematic but *empirical* way with particular microstructural variables such as grain size. A control of grain size could in principle be achieved in a pure single-phase polycrystal by suitable heat treatments. But in real materials even small amounts of impurity present will be distributed differentially between grain interiors and grain boundaries. Thus the chemical and physical nature of the grain boundaries will tend to change with grain size. Where detailed data do exist for ranges of related materials the results are often not capable of rational explanation (Davidge, 1974; Pratt, 1975). Sometimes surface energy increases with increase in grain size, sometimes it decreases and in other cases varies less systematically. Part of the difficulty must lie with the variation in the ratio of transgranular to intergranular fracture as the grain size varies. Figure 6.2 shows how the mode of fracture of a particular batch of MgO polycrystals changes from predominantly intergranular fracture at small grain size to predominantly transgranular fracture at large grain size. Some variation in surface energy might thus be expected on the basis that cleavage fracture consumes more energy than grain boundary fracture.

A comprehensive set of results for intercrystalline fracture has been obtained by Simpson (1974) for Al_2O_3 using notched beams. Materials with a range of porosity and grain size were produced by various heat treatments. An important (but fortuitous) feature of these materials is that fracture was almost entirely intercrystalline for all microstructures, so that one key variable, the mode of fracture, could be eliminated. This constancy of fracture mode was partly due to the situation of the porosity at grain boundaries. γ_i falls markedly with increase in porosity

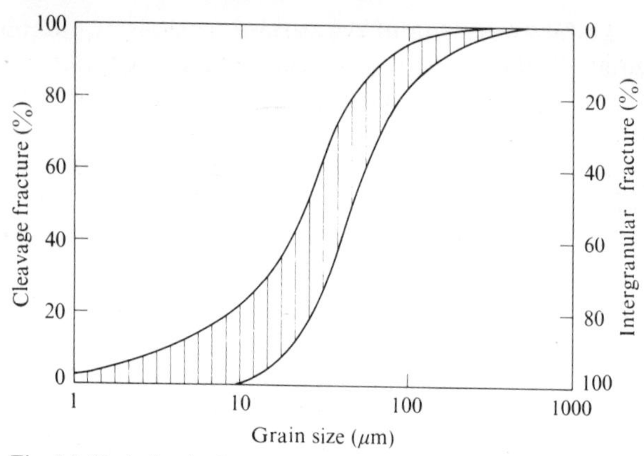

Fig. 6.2. Variation in the mode of fracture of MgO polycrystals with grain size. (After Rice, 1972.)

as expected; this is in agreement with the data of Evans and Tappin (1972) for similar aluminas. γ_i also falls with increase in grain size, fig. 6.3, and Simpson suggested that this may be due to the larger residual stress (or more correctly strain energy, see § 6.2) present in the large-grain-size material due to anisotropic thermal contractions in different crystallographic orientations on cooling after fabrication.

Some further generalisations are possible. First, the grain-boundary fracture energy γ_{gb} should be less than the cleavage energy: atomic bonding at grain boundaries is imperfect; impurities or grain anisotropy can lead to concentrations of strain energy at the boundaries; porosity if

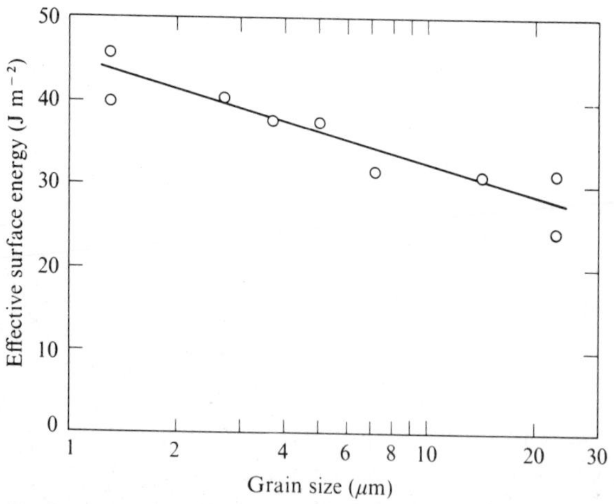

Fig. 6.3. Variation of γ_i with grain size for Al_2O_3 polycrystals. (After Simpson, 1974.)

present is often situated at grain boundaries. Secondly, γ_{gb} should be less at large grain size than at small grain size because the concentrations of impurity, porosity and strain energy are all greater at large grain size. Thus in materials where there is a transition from inter- to transcrystalline fracture with increase in grain size, γ_i should also increase and in material without a transition γ_i should decrease. In general this seems to be observed. This analysis is consistent with γ_p being the major contributing effect to γ_i. Fracture at grain boundaries occurs at lower stresses ($\gamma_{gb} < \gamma_0$) than for cleavage fracture and this will affect the γ_p term in sympathy.

In view of the numerous factors affecting γ_i it seems unlikely, in the current state of knowledge, that quantitative calculations of surface energy will be possible for polycrystalline materials. Instead it is best to regard γ_i as an empirically determined parameter.

6.2 Inherent flaws

For flaws of a given shape, it is the largest flaw that produces the greatest stress concentration. We thus need to identify the largest defect. The simplest inherent flaw to appreciate is the pore. Figure 6.4 shows a pore in a polycrystalline alumina revealed by scanning electron microscopy of a fracture face. The pore is much larger than the surrounding grains and of particular significance are the sharp cusps around the surface of the pore which partly penetrate between neighbouring grains. For a pure

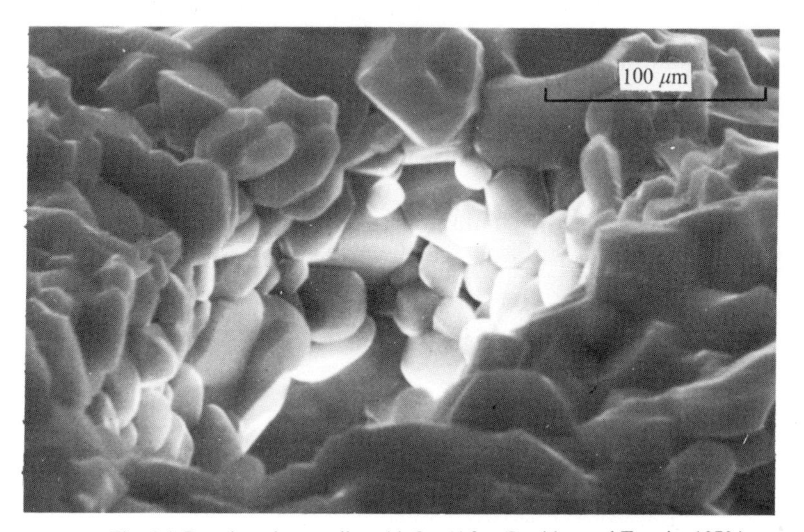

Fig. 6.4. Pore in polycrystalline Al_2O_3. (After Davidge and Tappin, 1970.)

material the configuration at the grain boundary would be of the type shown in fig. 1.8 with the flaw atomically sharp. With second phases present, however, the tip of the cusp could be somewhat rounded and the stress-concentrating effect correspondingly reduced. At the other extreme, pores in glass are likely to be perfectly spherical in shape because this is the minimum-energy configuration. In this case the stress-concentrating effect of the flaw will raise the applied stress only by a factor of three. It is, therefore, both the largest dimension of the pore and also the smallest radius of curvature at the pore surface that are important in determining the stress to propagate fracture from the pore.

Apart from pores, all ceramics are likely to contain other forms of defect near the specimen surface, for a variety of reasons. Exposure to high temperature during fabrication results in grain-boundary grooving; the grooves concentrate stresses along the boundaries which have a low energy requirement for fracture. (This particular effect is put to good use in thermal-etching techniques where specimens are heated, after polishing, to high temperatures at which the grain boundaries are revealed due to thermal grooving.) Many materials require machining operations after firing and this can lead to a surface-damaged layer. Finally, there is always the possibility of accidental damage to the surface of the material in use. Figure 6.5 illustrates a typical configuration of grain boundaries near a ceramic surface together with the type of crack expected from modest machining damage. Two types of cracks are shown: those propagating along the grain boundaries and those along cleavage planes. The sketch shows that in both cases the cracks have propagated for approximately one grain dimension. At this point there is a natural barrier to further propagation in that the grain-boundary crack needs to change direction considerably, whereas the cleavage crack has either to nucleate a new cleavage crack in the next grain, which will not be coplanar with the original crack, or alternatively nucleate a new crack along a grain boundary. It is expected, therefore, and commonly observed, that machining cracks penetrate approximately a grain dimension from the surface.

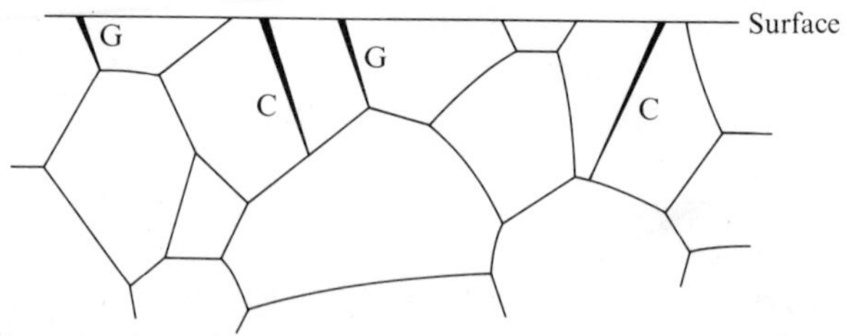

Fig. 6.5. Grain-boundary cracks (G) and cleavage cracks (C) near the surface of a ceramic and limited approximately to a grain dimension.

Cracks can also form at internal grain boundaries. In anisotropic materials such as Al_2O_3 or BeO, which are of hexagonal structure, the thermal expansion coefficients vary with crystal orientation. Consider the two grains shown in fig. 6.6 which are assumed to have their respective *a* and *c* crystallographic axes parallel to the plane of the boundary. For both beryllia and alumina the *c*-axis expansion is slightly greater than the *a*-axis expansion by $\sim 1 \times 10^{-6}\ K^{-1}$ $(=\Delta\alpha)$. As these materials cool down from the fabrication temperature the differing

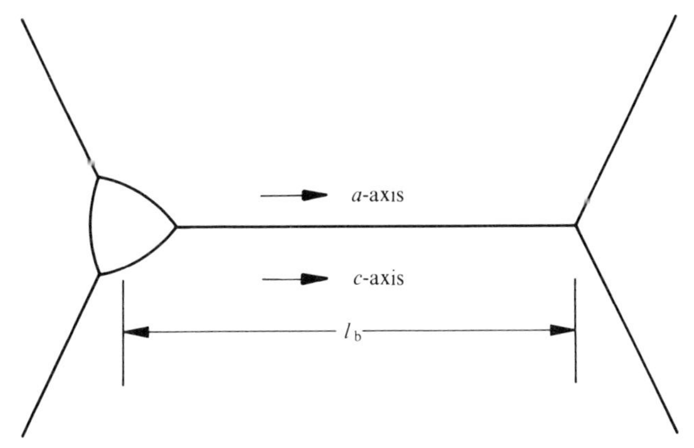

Fig. 6.6 Adjacent grains with pore at triple junction.

contraction rates in the two orientations in the adjacent grains cause strains to build up near the grain boundaries increasingly as the temperature falls. If it is assumed arbitrarily that no plastic relaxation is possible at temperatures < 1020 °C then the elastic strains are built up over a temperature range (ΔT) of 1000 K on cooling to 20 °C. The differential strain set up near the boundary is thus $\varepsilon_{gb} = \Delta\alpha\Delta T/(1-\nu) \sim 1.3 \times 10^{-3}$. Whether or not this leads to fracture of the grain boundary depends on the usual two factors: the theoretical strength must be exceeded at some point, and it must be energetically favourable for a crack to grow. If, for example, a pore is situated at one of the triple grain junctions, as shown in fig. 6.6, this can provide the necessary stress concentration to nucleate a crack – either spontaneously due to the above internal strain energy, or with the additional assistance of an external stress.

Note that the strain and thus the stress σ_{gb} at the boundary does not include a term for the length of the boundary. The strain energy available, however, depends on the volume of the strained region near the grain boundary. If this is assumed to have a roughly spherical shape with the common grain-boundary area as the diameter, then the strain energy will vary roughly as l_b^3 where l_b is the length of the common boundary. On the other hand, the energy required to fracture the boundary completely will vary only as l_b^2. Thus, as the size of the common grain-boundary area increases, there will be an increasing tendency to satisfy the energy requirement for fracture at the boundary.

Clarke (1964) estimates that the rate of release of internal strain energy per unit depth of boundary $(\mathrm{d}U_i/\mathrm{d}C)$ as a crack of length C grows along the boundary is given by

$$\frac{\mathrm{d}U_i}{\mathrm{d}C}=\frac{\sigma_{gb}^2(l_b-C)}{12E(1-\nu^2)}. \tag{6.3}$$

and that if C is assumed to be $l_b/5$, then spontaneous fracture of the boundary should occur when

$$\varepsilon_{gb}\sim\left(\frac{24\gamma_{gb}}{El_b}\right)^{1/2}. \tag{6.4}$$

This is consistent with the observation for beryllia that grain-size cracks are observed for a grain size >120 μm. Although (6.3) shows that $(\mathrm{d}U_i/\mathrm{d}C)$ decreases as C increases, it is likely that U_i is sufficiently high for the crack to propagate for a length l_b once fracture is initiated.

Additional evidence comes from studies of neutron-irradiated beryllia (Davidge and Tappin, 1968*b*). During neutron irradiation the displaced atoms tend to agglomerate as dislocation loops predominantly parallel to the basal planes in the structure. This means that the *c*-axis expands at a greater rate than the *a*-axis. This effect is thus additive to that mentioned for anisotropic thermal expansion and grain-boundary cracks are observed in irradiated beryllia of a given grain size after a particular critical neutron dose. It is found that the neutron dose required to produce cracks increases as the grain size diminishes, again in accordance with the argument developed above. Figure 6.7 shows grain-boundary cracks in beryllia irradiated to a fast-neutron dose of 7×10^{23} m^{-2}. An additional interesting observation is that the strength of material with single-grain-size cracks is no different from that of the as-fabricated material. A crack can thus propagate relatively easily for one grain dimension but then requires a much higher surface-energy value and applied stress to propagate further.

Similar arguments apply to polyphase materials where the various phases have differing expansion coefficients and additionally different elastic properties. A number of important principles have been established from work on model systems of ceramic particles in glassy matrices. Such systems have many attractions. The matrix can be regarded as structureless, apart from inherent small cracks typically

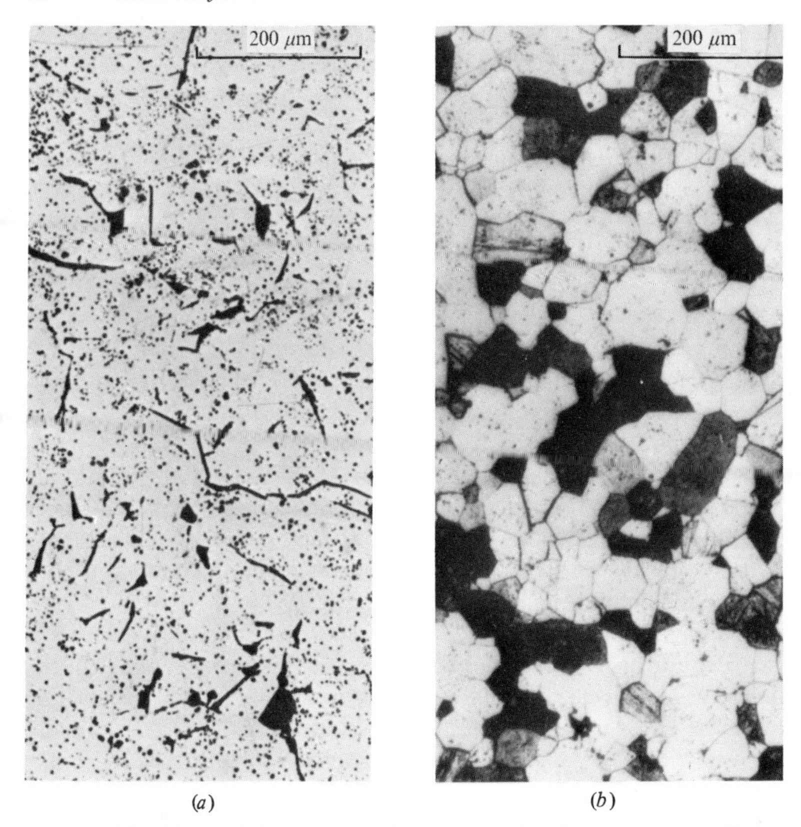

Fig. 6.7. Grain-boundary cracks in neutron-irradiated BeO: (*a*) polished; (*b*) etched to reveal individual grains. (After Davidge and Tappin, 1968*b*.)

50 μm long, and the physical interactions between matrix and particle that lead to crack formation can be studied in some detail. Specimens are easily fabricated by hot-pressing techniques. The types of available glasses and particles are numerous so that the effects of large variations in physical properties can be studied.

It is found that the important strength-controlling factors include: (*a*) the thermal expansion coefficients of the two phases; (*b*) the elastic properties of the two phases; (*c*) the volume fraction of the crystalline phase; and (*d*) the particle size of the crystalline phase.

The theory concerning the stress system around particles in an isotropic medium due to a difference in expansion coefficients is well established. Stresses are set up within and around the particles as the body cools down from the fabrication temperature. A spherical particle will be subjected to a pressure P (equivalent to radial and tangential stresses of $-P$), and the matrix to radial and tangential stresses of $-PR^3/r^3$ and $+PR^3/2r^3$ respectively, where R is the particle radius and r the distance from a point in the matrix to the centre of the particle. Ideally, these equations hold only for the case of a single particle in an infinite isotropic matrix, but they represent a satisfactory approximation at low particle concentrations. It can be shown that

$$P = \frac{\Delta\alpha\Delta T}{(1+\nu_m)/2E_m + (1-2\nu_p)/E_p} \tag{6.5}$$

where $\Delta\alpha$ is the difference in the two expansion coefficients, ΔT is the cooling range over which the matrix plasticity is negligible (taken to be from the annealing-point temperature to ambient), and $\nu_{m,p}$, $E_{m,p}$ the Poisson's ratio and Young's modulus of the matrix and particle.

Equation (6.5) again shows that the magnitude of the stresses is independent of particle size. Experimentally it is observed that for a particular system cracking occurs only around particles greater than a critical size. The formation of cracks must depend, therefore, on both the stress magnitude and the particle size.

The nature of the cracking, if it occurs, depends on whether the particles contract more or less than the matrix during cooling. In the former case P is negative and cracking is circumferential around the particles. In the latter case P is positive and cracking occurs radially from the particles; this is more deleterious to strength because cracks from individual particles can easily link up. These two cases are shown in fig. 6.8 with specific examples in model materials of glasses, containing 0.10 volume fraction of ThO_2 spheres with graded diameters (Davidge and Green, 1968). In materials with $\alpha_p > \alpha_m$, circumferential cracks are observed in as-fabricated specimens, but only around spheres larger than a critical diameter which depends on $\Delta\alpha$. The argument is thus similar to that developed for the anisotropic materials discussed earlier. The critical diameter can be estimated approximately by an energy-balance criterion. The elastic energy released from the sphere and neighbouring matrix is equated to the surface energy required to form the observed, roughly hemispherical, crack. In specimens containing spheres smaller than the critical diameter, cracks appear at stresses below the macroscopic fracture stress so that the same weakening effect as in a cracked specimen should occur. For the two glasses studied, there is a good quantitative agreement between strength and particle size using a crack size equal to the sphere diameter in the Griffith equation. In compacts where $\alpha_p < \alpha_m$ very low strengths are obtained because the radial cracks produced readily link together. Thus, the main effect when

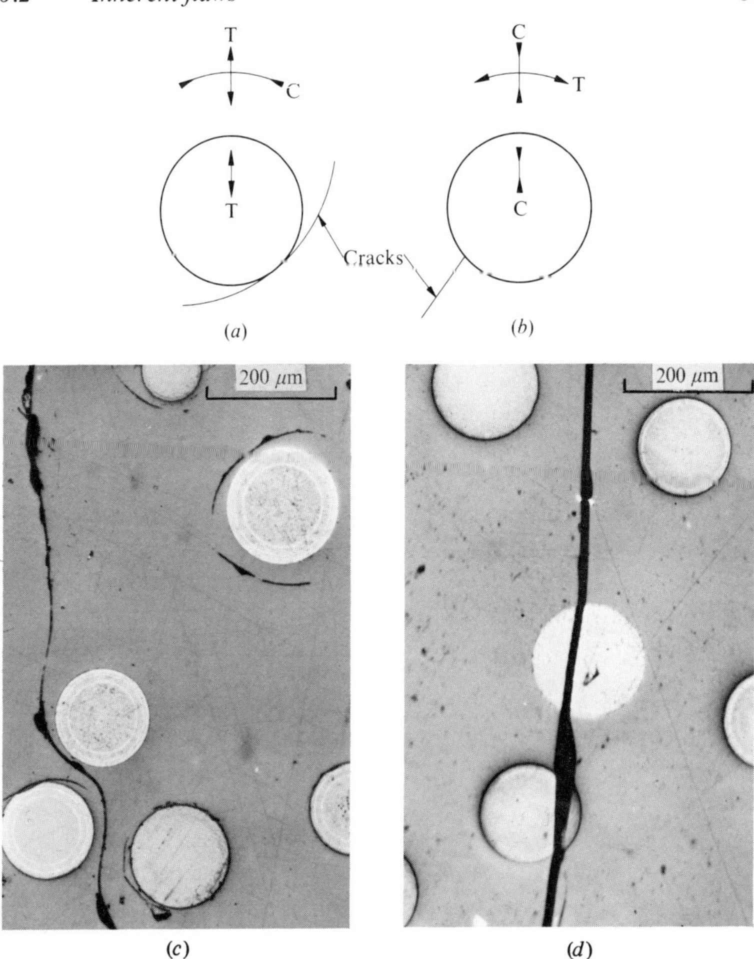

Fig. 6.8. Stress distributions and cracks around particles when (*a*) $\alpha_p > \alpha_m$ and (*b*) $\alpha_p < \alpha_m$. Observed fracture paths in ThO_2-sphere-in-glass composites for (*c*) $\alpha_p > \alpha_m$ and (*d*) $\alpha_p \sim \alpha_m$. (After Davidge and Green, 1968.)

α_p and α_m are unequal is to introduce cracks larger than the inherent Griffith flaws; this leads to a reduction in strength.

A parallel case of industrial significance is that of electrical porcelain. This is a typical three-component system, as described in § 1.3, with the filler particles generally comprised of quartz. During cooling after fabrication the quartz particles contract much more than the matrix due to a phase transformation in the crystalline quartz. When the quartz particles are $\gtrsim 100$ μm, cracks are produced circumferentially around the particles. This is demonstrated in fig. 6.9 where both etched and polished

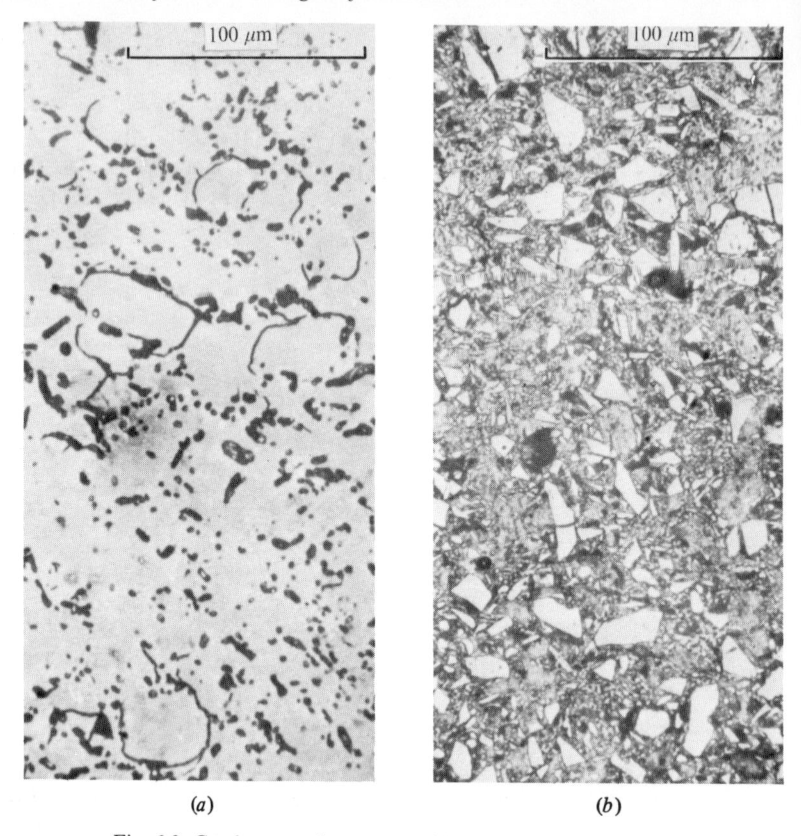

Fig. 6.9. Cracks around quartz particles in electrical porcelain: (*a*) polished; (*b*) etched to reveal quartz particles.

sections are shown. (It is difficult to reveal both the quartz grains and the cracks around them in the same micrograph.) The solution to making stronger electrical porcelains thus lies in using quartz particles of a finer size or, alternatively, in using a different filler, such as alumina, which is more compatible with the matrix from an expansion point of view.

The stresses due to differences in elastic constants have been considered by Hasselman and Fulrath (1967) for alumina spheres in glass. Stresses occur only in the presence of an applied stress and the important point is that the applied stresses become magnified in the vicinity of the sphere. Under a uniaxial tensile stress, the region of significant stress magnification is limited to a zone around the direction of the tensile stress. This is illustrated in fig. 6.10. The maximum stress concentration is typically 1.4 times in this system and occurs at a point on the surface of the sphere. The hatched area shows the region subjected to a stress concentration of more than 1.2 times. Again the magnitude of the stress is independent of the sphere diameter. This is a much less drastic effect than that of thermal-expansion mismatch and is less likely to lead to cracking.

Experimentally, systems containing spheres generally $>10\ \mu m$ diameter have been examined. Hasselman and Fulrath (1966) studied alumina spheres 15–60 μm in diameter in a glass with an expansion

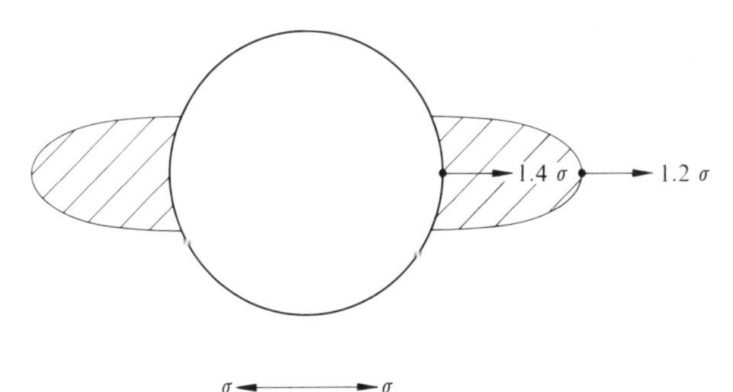

Fig. 6.10. Stress magnification due to differences in elastic properties of particles and matrix. The hatched areas are subjected to stress magnification from 1.2 to 1.4 times for alumina particles in glass. (After Hasselman and Fulrath, 1967.)

coefficient matched to that of alumina. The volume fraction of alumina was varied up to 0.475. For a given size of sphere the strength was essentially constant with increasing volume fraction of alumina up to a critical value and then increased. The critical value corresponded to a mean sphere spacing of $\sim 40\ \mu m$ which was the Griffith flaw size of the plain glass. The strength of the composites with high volume fractions of alumina was consistent with a Griffith crack size equal to the mean sphere spacing. In these systems differences in expansion coefficient were eliminated and the stress-concentration effect due to differences in elastic properties appeared to have no significant effect. For high volume fractions of small particles with an expansion coefficient matched to the glass, a useful strengthening effect is thus obtained by limiting the Griffith crack size in the matrix.

The above discussion has demonstrated that the main flaws in ceramics range from pores to grain and phase boundaries. Whether pores or boundaries are the most important defect clearly depends on their relative sizes. Figure 6.11 sketches some idealised microstructures to emphasise the relative importance of pores and grains. The four parts represent a fully dense material and porous materials where the pores are either smaller than, similar to or greater than the size of the grains. In the first two cases pores should play a minor role, an intermediate role in the third case and a predominant role in the fourth case. This is borne out by experimental data as we shall see later.

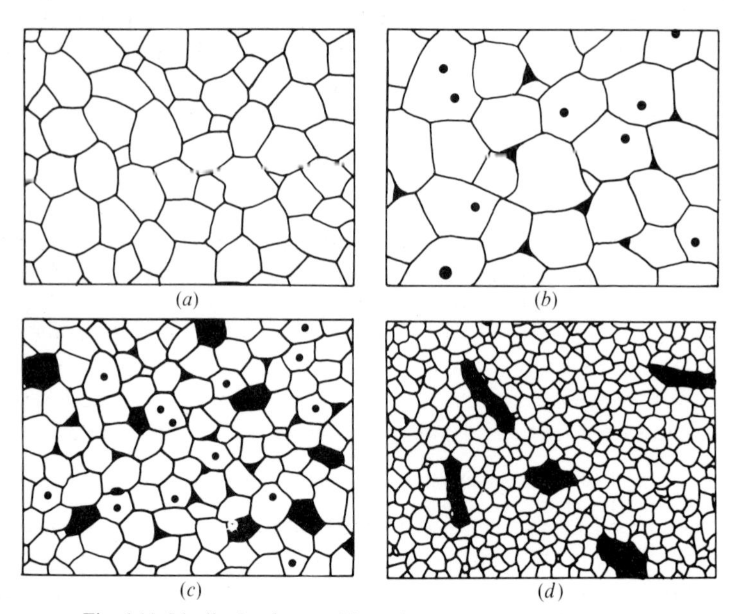

Fig. 6.11. Idealised polycrystalline microstructures: (*a*) fully dense; (*b*) porous, with small pores; (*c*) porous, with pores and grains of similar size; (*d*) porous with large pores. (After Davidge, 1972.)

Finally, we recall that cracks can be generated by several plastic deformation processes. We discussed in chapter 4 the production of microcracks through dislocation motion and pile-ups at various obstacles such as slip bands or grain boundaries. The plasticity associated with grain boundaries leading to crack formation during creep at high temperatures was mentioned in chapter 5.

6.3 Temperature dependence of the strength of selected engineering ceramics

A few comments are relevant before we discuss detailed behaviour for particular materials. The general pattern follows that shown in fig. 1.13 although some materials, such as reaction-bonded silicon nitride, show only a region *A* behaviour. At the other extreme uranium dioxide shows all three regions of behaviour. The majority of materials show region *A* and *B* behaviour but there is a large variation in the temperature separating these regions: ~20 °C for MgO and ~1400 °C for self-bonded SiC. The factors controlling this transition temperature can be attributed to many causes: dislocation motion in MgO and UO_2; grain-boundary sliding associated with second phases or impurities in Al_2O_3 and hot-pressed silicon nitride; melting of the silicon phase in self-bonded SiC. In general the more ionic materials, for example MgO, show less refractory behaviour than the more covalent materials, for example Si_3N_4, although this is greatly affected by the presence of impurities.

Data for the strength of a number of ceramics including the temperature dependence are now discussed to illustrate how the major principles described so far can be used to generate a quantitative understanding of the strength of ceramics from the materials-science aspect. The discussion is restricted to four materials – magnesia, uranium oxide, reaction-bonded silicon nitride and alumina – rather than presenting an exhaustive review of available current information. An interpretation of data for other materials should follow the same broad principles.

Magnesium oxide

Magnesia is an unusual ceramic in that plastic deformation is possible at ambient temperature. The fracture-initiating flaws can thus be either the inherent flaws, or those produced by dislocation flow. Which of these flaws is the more important depends on the state of the specimen surface, the flow stress of the material and its grain size. Figure 6.12 shows bend-strength v. grain-size data for two batches of the same polycrystalline magnesia with different surface conditions: machined and chemically polished (Evans and Davidge, 1969*a*). Measurements of the surface energy to propagate cracks beyond one grain size indicate a value 5 J m^{-2}, which is relatively insensitive to grain size. The machined specimens are observed to have grain-sized cracks over their surfaces and the strength of these materials agrees well with the prediction from (3.17) with $\gamma_i = 5$ J m^{-2} and C equivalent to the grain size. The second set of specimens is chemically polished after machining to remove all surface damage. The strength of these specimens is considerably greater than that of the machined specimens, particularly at the larger grain sizes. The chemical polishing also produces grooves along grain boundaries, as in fig. 1.8, and these should be atomically sharp. The maximum groove depth is 3 μm and the stress to propagate a crack from the groove is estimated to be 275 MN m^{-2}, using $\gamma_i = 0.75$ J m^{-2}. This is greater than the observed strengths and thus, in this case, propagation of cracks from grain-boundary grooves is not the critical mechanism. The grain-size dependence of strength for the polished specimens is small and the data were found to extrapolate at infinite grain size to the dislocation flow stress of the material σ_0. It is believed that the critical event for these specimens is the initiation of cracks at grain boundaries following a pile-up of dislocations at the boundary. This mechanism produces grain-sized cracks which can then propagate at the lower stress indicated by the data for the machined specimens. The strength of the polished crystals thus follows (4.4), and the fracture stress is controlled by the stress to nucleate the crack.

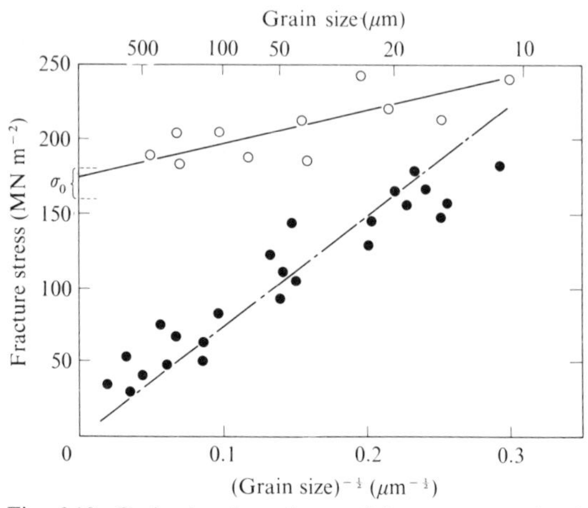

Fig. 6.12. Grain-size dependence of fracture strength of MgO at 20 °C. ○, polished; ●, machined surfaces. (After Evans and Davidge, 1969*a*.)

The temperature dependence of strength for MgO of 25 μm grain size with a polished surface is shown in fig. 6.13 (Evans *et al.*, 1970). The fracture stress falls steadily with increase in temperature and reaches a constant value at ~1200 °C. Measurements as a function of temperature are shown of the flow stress of single crystals of similar purity, and also the compressive flow stress, as estimated from an axial compressive test, of polycrystals of grain size 25 μm but only at temperatures >1100 °C because brittle fracture intervened at lower temperatures. The fracture stress exceeds the single-crystal flow stress by ~50 MN m^{-2} at 20 °C with this difference at first increasing and then decreasing as the temperature increases. The factor k^* in (4.4) should thus vary likewise. The increase in k^* over the lower temperature range has been attributed to the tendency for slip bands to become more diffuse with increase in temperature so that the stress-concentration effect at the head of the slip band is diminished. At the higher temperatures the additional possibility of grain-boundary sliding is a complicating factor which could either extend or blunt the cracks, and detailed analysis is not available.

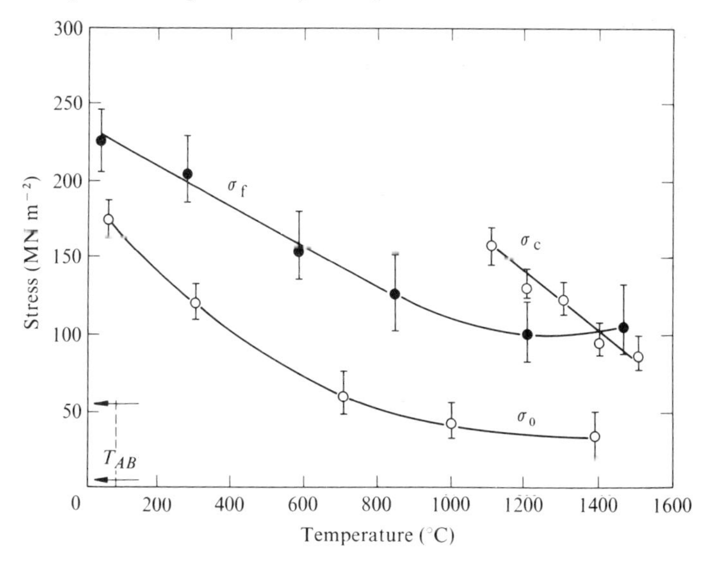

Fig. 6.13. Temperature dependence of the fracture stress σ_f and compressive flow stress σ_c for polished specimens of 25 μm-grain-size MgO, compared with the flow stress σ_0, of single crystals of similar purity. (After Evans *et al.*, 1970.)

Uranium dioxide

The temperature dependence of the strength of stoichiometric UO_2 is of particular interest in that all three regions of behaviour are observed. Data are shown in fig. 6.14 including values for the flow stress of the material, as measured in compressive tests, plus the displacement at fracture when plastic deformation is observed at temperatures > 1200 °C (Evans and Davidge, 1969*b*). No data are available for the single-crystal flow stress.

It is evident that the fracture process can be separated into the three distinct temperature regions, indicated in fig. 1.13: region *A* where the fracture stress is substantially lower than the flow stress and fracture occurs in a brittle fashion; region *B* where the fracture and flow stresses are similar although fracture is still brittle; region *C* where the fracture stress is in excess of the flow stress and fracture is ductile.

Since the fracture stress in region *A* is always lower than the flow stress, fracture occurs either by the extension of the pre-existing flaws or is initiated by limited plastic flow. To determine which of these processes operates one must calculate the stress to extend the pre-existing flaws. If this is similar to the observed fracture stress, fracture occurs by the extension of the flaws; if it is larger, fracture initiated by plastic flow is likely.

The biggest pre-existing flaws are large pores $\sim 50\ \mu m$. Where a grain boundary emerges at one of these pores a grain-boundary groove with a sharp angle is developed. The pore can extend to a length equivalent to the pore size plus a grain size at a stress lower than the observed fracture stress because of the low surface-energy requirement for grain-boundary

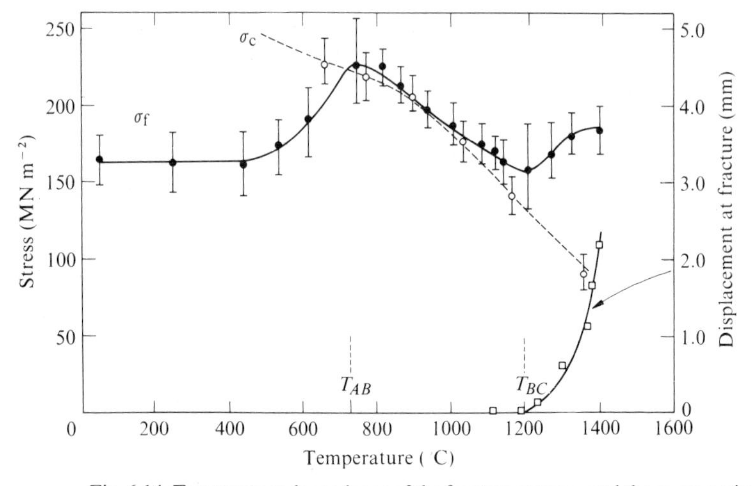

Fig. 6.14. Temperature dependence of the fracture stress σ_f and the compressive flow stress σ_c for stoichiometric UO_2, grain size 8 μm. A displacement at fracture, in the bend test, is detected in region *C* at temperatures > 1200 °C (as indicated by the right-hand scale). (After Evans and Davidge, 1969*b*.)

fracture. The stress to extend the largest pores at room temperature is 165 MN m^{-2}. This is similar to the observed fracture stresses at this temperature, indicating that fracture occurs by the extension of the large pores. The variation of the pore extension stress with temperature may be ascertained using γ_i values measured at the relevant temperature. γ_i increases with temperature from 8 J m^{-2} at 20 °C to 15 J m^{-2} at 600 °C. In this temperature range there is good agreement between the pore-extension stress and the measured fracture stress.

In region *B* fracture occurs at the macroscopic compressive flow stress. The fracture stresses are also well below the stresses to extend the pre-existing flaws. Grain-sized microcracks are formed in this temperature region. If a number of such cracks, in the vicinity of a pore, link up then catastrophic fracture should occur at a stress near the flow stress. The role of plastic flow is thus to extend the pre-existing flaws, prior to catastrophic propagation.

It is not clear in UO_2 how grain-size microcracks are formed at the flow stress. They could be nucleated either by dislocation motion or grain-boundary sliding. Grain-boundary sliding is not usually observed, however, in stoichiometric UO_2 at these temperatures. The precise dislocation mechanism must be more complex than that for polycrystalline MgO where cracks form at stresses below the macroscopic flow stress.

In region *C* fracture is preceded by general plastic flow which must involve additional slip on the secondary slip system (table 4.2). In this way crack blunting is also possible and the fracture stress rises slightly with increase in temperature.

Reaction-bonded silicon nitride (RBSN)

Strength data as a function of temperature for a typical material of density 2.55 Mg m^{-3} are shown in fig. 6.15 (Evans and Davidge, 1970). When tested in argon, RBSN shows strengths which fall only very slightly with increase in temperature; there is no evidence of any plastic effects in these short-term tests even at the highest temperature used, 1800 °C. Although dislocations have been observed (Evans and Sharp, 1971) by electron microscopy in material deformed at high temperature, the deformation was under very high compressive stresses. Dislocation movement is thus not considered to be an important factor in determining the high-temperature strength of RBSN.

The most obvious defects in RBSN are pores, fig. 1.6. The porosity results from the specific fabrication route and is generally $>15\%$ and typically 20–25%. Such high porosities are necessary to allow the penetration of nitrogen to the centre of the article being nitrided to give a high conversion of silicon to silicon nitride and porosity is thus mainly open-type. The pore size has a wide distribution but the majority of pore volume is concentrated in very fine pores, typically 80% <0.1 μm. The maximum pore sizes observed are in the range 20–30 μm, and although these are few in number they have a dominant effect on the strength. Small maximum pore sizes are generally associated with fine-particle-size silicon, high fired density, nitriding at temperatures below the melting point of silicon and with static rather than flowing atmospheres.

Both the α- and β-forms of silicon nitride are generally detected in RBSN. The relative amounts of these phases vary widely between different materials, but generally α-Si_3N_4 formation tends to be favoured by low nitriding temperatures and β-Si_3N_4 by high nitriding temperatures. Electron microscopy shows that the grain size is very fine. Individual grains <5 μm are observed but these are surrounded by a very fine grain sized material $\ll 1$ μm. Electron diffraction reveals a tendency for the larger grains to be β-Si_3N_4 and the smaller grains α-Si_3N_4. Many pores contain fine fibres of α-Si_3N_4.

The largest pore size rather than the largest grain size is thus expected to control the strength of RBSN. Substituting values for the largest pore size into (3.17) with the measured surface-energy values enables comparisons to be made between calculated and observed fracture stresses. Data for materials from various origins show that very good agreement

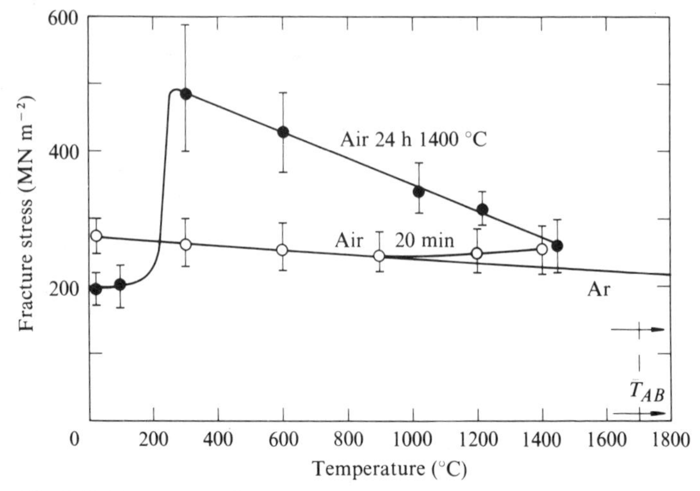

Fig. 6.15. Temperature dependence of the fracture stress of reaction-bonded silicon nitride tested in Ar (lower straight line); in air after holding for 20 min (○) (coincident with data in argon for temperatures <900 °C); and in air after holding in air for 24 h at 1400 °C and cooling to test temperature (●). (After Evans and Davidge, 1970.)

is obtained. There is no significant variation in the surface energy with temperature and thus the very small temperature dependence of strength is readily understood.

The high-temperature mechanical properties of RBSN in oxidising environments are dominated by the effects of oxidation which are of particular significance for a porous material (Davidge *et al.*, 1972). This is of great importance because many of the envisaged applications for RBSN involve operation in oxidising atmospheres. Oxidation is quite rapid at temperatures as low as 1000 °C; for example, 100 h in air gives a 5% weight gain, equivalent to a silica content of 14%. Although the rate of the oxidation reaction increases with increase in temperature the total amount of oxidation of a specimen does not. Much greater weight gains are found on oxidation at 1000 °C than at 1400 °C. This apparent anomaly depends on the relative fluidity of the oxidation product at the

(*a*)

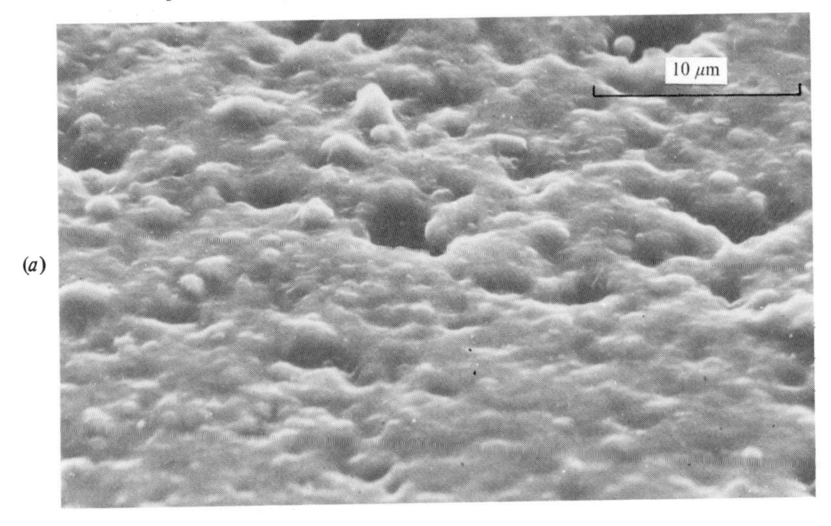

(*b*)

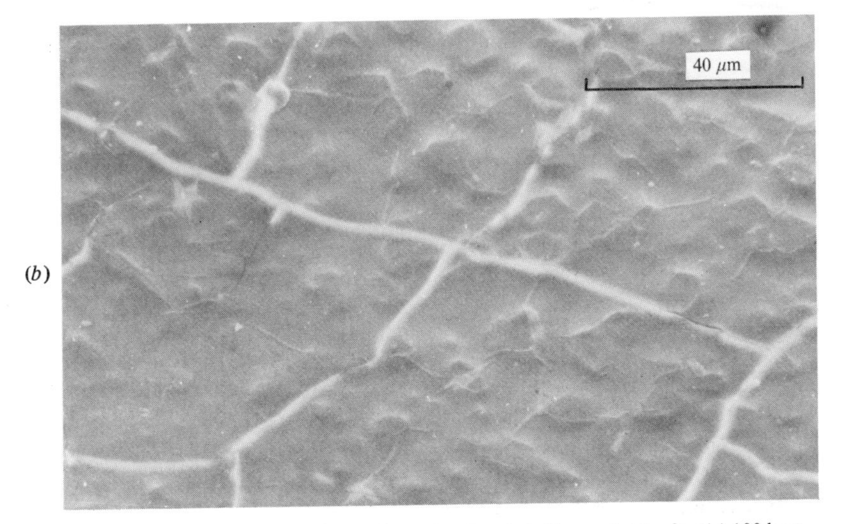

Fig. 6.16. Oxidised surfaces of reaction-bonded silicon nitride after (*a*) 100 h at 1000 °C and (*b*) 24 h at 1400 °C (note the crazed surface). (After Davidge *et al.*, 1972.)

various temperatures. Figure 6.16 shows surfaces of RBSN oxidised at 1000 and 1400 °C. At 1000 °C the silica is not very fluid and the surface pores are still obvious. At 1400 °C, however, the silica is very fluid and after very short oxidation times the surface of the specimen becomes covered by a continuous film. This acts as a barrier to further internal oxidation and the oxidation effects become concentrated at the surface.

Another important difference between oxidation at these two temperatures is that the oxidised surface at the higher temperature contains a network of cracks after cooling to ambient temperatures. The main oxidation product is cristobalite, and this has a large thermal-expansion mismatch with RBSN on cooling through temperatures ~270 °C where a phase transition from β- to α-cristobalite takes place. Cracks have not been observed in material oxidised at 1000 °C where the layers of silica are much thinner.

When tests are conducted in air, the strength at temperatures >1000 °C is slightly enhanced compared with the values obtained in argon. Very pronounced effects on strength are obtained by oxidising specimens at 1400 °C for 24 h and then cooling directly to the test temperature. The results in fig. 6.15 show that the strength increases markedly to 500 MN m^{-2} as the test temperature after oxidation is reduced. Here the surface layer is believed to act in the manner of a conventional glaze whereby it develops a compressive stress due to a lower thermal expansion coefficient. However, once the specimens are cooled through the cristobalite transformation temperature the strength drops to a value lower than that for the original material. In this case the specimen surface is highly crazed as shown in fig. 6.16. Once having been cooled to ambient temperature the material then exhibits low strength values if the temperature is again raised. Oxidation at 1000 °C produces quite different effects and strength is enhanced by ~25%, irrespective of whether the material has first been cooled to ambient temperature or not. Oxidation at 1000 °C is thus beneficial and has been related to the rounding of pores by oxidation.

Alumina

Of the engineering ceramics discussed here alumina is probably the most important in that it exists over a wide range of formulations from nominally pure material used, for example, as furnace tubes or envelopes for lamps, to less pure forms containing typically ~10% of a mixed silicate phase used, for example, in spark plugs. In addition, aluminas have long been used as model materials and there is extensive information available on their mechanical properties. In this section we discuss two aspects in some detail: the strength of material at ambient temperatures and the temperature dependence of typical pure and impure materials.

In many cases the strength can be understood in terms of the stress to propagate flaws as summarised in fig. 6.11. For two particular commercial aluminas, Davidge and Tappin (1970) observed that the maximum grain size and pore size were approximately equal and that the mean strength correlated with the stress to propagate a flaw equal in size to the sum of the largest pore size and the largest grain size.

In materials with higher porosity the situation is more complicated in that pores may link together prior to catastrophic failure so that the description of strength in terms of the extension of an individual flaw is inadequate. Evans and Tappin (1972) prepared a range of experimental aluminas with controlled amounts of porosity. When the separation between the pores is large compared with their size, the critical flaw extension stress is in accordance with the models described in § 3.7 for the extension of individual flaws; for example, at 10% porosity the pore

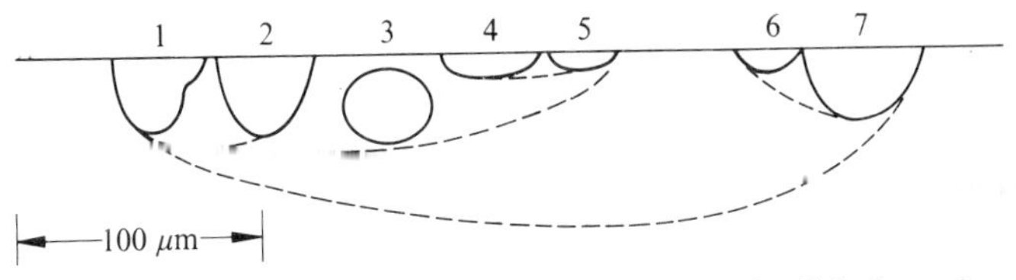

Fig. 6.17. An array of surface pores (in a 95% alumina) that linked together at sub-critical stresses to form a critical flaw. (After Meredith and Pratt, 1975.)

spacing is such that the possibility of flaw linking prior to catastrophic failure is small. On the other hand, in materials with 50% porosity a sub-surface flaw is found to link up with a surface flaw at a stress ~75% of the stress to extend the isolated surface flaw, in good agreement with calculations.

In a number of commercial aluminas containing 10–20% porosity, Meredith and Pratt (1975) performed some careful experiments whereby the fracture origins of individual specimens were identified by examination of the fractured faces of the specimens. Figure 6.17 shows the distribution of pores (numbered 1–7) near the failure origin of a particular specimen with a fracture stress of 342 MN m^{-2}. It was demonstrated by application of (3.25) that the pores 4 and 5 would link together at a stress ~318 MN m^{-2}, at which stress the larger pores 1 and 2 would also have linked. These can then link with pore 3 at the same stress to produce a large semi-elliptical crack of depth 40 μm and length 204 μm. The stress to propagate this larger flaw, from (3.24), is 435 MN m^{-2} and thus is still significantly in excess of the observed fracture stress. A final linking between this flaw and pores 6 and 7 is calculated to occur at a stress of 339 MN m^{-2} to produce a flaw of depth 70 μm and length 330 μm. The stress to propagate this flaw is virtually the same as the previous linking stress, which is also the observed fracture stress of the material, and it is thus not possible to identify unambiguously which of these latter two events leads to the ultimate catastrophic failure. This exercise was repeated for other specific cases and demonstrates that the strength of individual specimens can be understood quantitatively provided that sufficient detail is known about the distribution of flaws near the fracture origin.

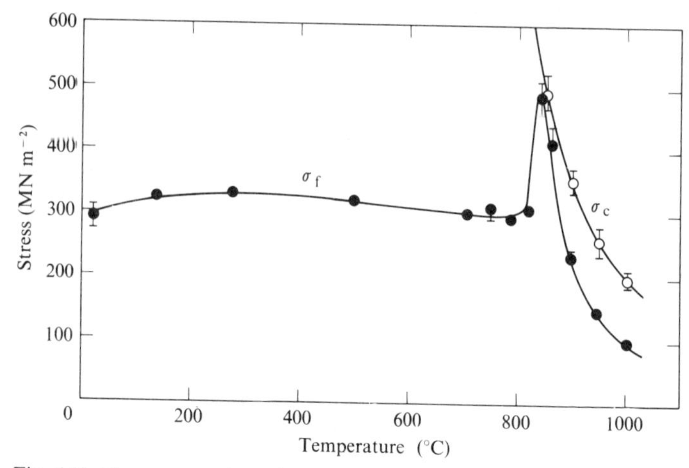

Fig. 6.18. Temperature dependence of the fracture stress σ_f and compressive flow stress σ_c of a 95% alumina (similar to that in fig. 1.9). (After Meredith *et al.*, 1975.)

Turning now to the effects of temperature for alumina, the strength of materials of nominal purity usually shows little variation at temperatures <1000 °C. A gradual fall-off in strength is observed at higher temperatures but this occurs at stresses much lower than those required to produce plastic deformation by dislocations. It is believed that the fracture strength here is controlled by a grain-boundary-sliding mechanism.

Particularly fascinating effects are found in aluminas containing $\sim 10\%$ of glassy phases and fig. 6.18 shows data. At temperatures $\lesssim 800$ °C the strength is fairly invariant with temperature. Over a short temperature range (<100 K) above this there is a very sharp peak in strength, the maximum strength enhancement being 50%. At higher temperatures the strength then falls very rapidly with further increase in temperature. Also shown is the flow stress of the material as measured by compressive tests. Plastic deformation in this material is first observed at temperatures >800 °C and the compressive flow stress falls very rapidly as the temperature is increased above 800 °C. This sharp increase in strength is associated with plastic effects in the glassy phase. It is believed that near the peak in strength the temperature is such that the viscosity of the glass permits a localised reduction of the stress near the tips of the critical cracks and effectively increases the surface energy through a term which increases the radius of curvature at the crack tip (see (3.13)). At slightly higher temperatures, however, the usual grain-boundary-sliding effects, assisted by the plastic second phase, nucleate or extend flaws and the strength is reduced.

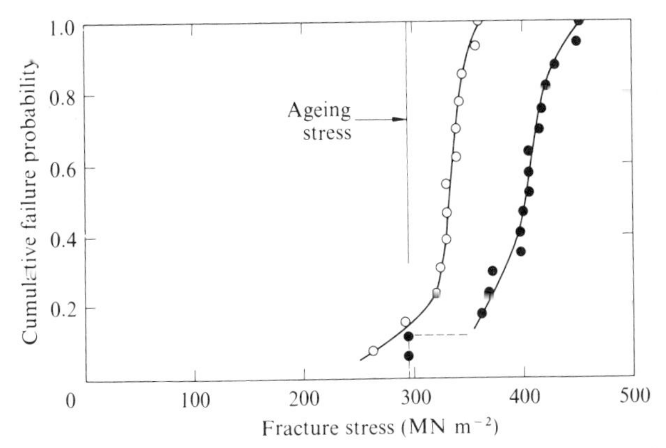

Fig. 6.19. The effect of ageing under stress at 650 °C on the strength of a 95% alumina tested at 650 °C. The unaged specimens (○) show a spread of strength values. The aged specimens (●) were held at 295 MN m^{-2} for 30 min and then fractured. Apart from two specimens that failed at the ageing stress the remainder are significantly stronger but with a similar spread of strengths. (After McLaren and Davidge, 1975.)

A related strength enhancement is that of ageing under stresses near to the fracture stress and data are shown for a similar alumina in fig. 6.19. A series of specimens were aged under a stress in the lower regions of the statistical spread of strength for unaged specimens, at a temperature ~650 °C. (Statistical variations in strength are discussed in § 9.1.) Although a small number of specimens break in short times under this stress the majority do not and, on testing to failure, by increasing the stress above the ageing stress, it is demonstrated that an enhancement of strength ~20% is obtained for ageing times > 1 h. An idealised model to explain this is presented in fig. 6.20 for a two-dimensional microstructure comprising hexagonal grains with triangular regions of a softer phase at the three-grain junctions. Fracture nucleates at a surface pore which can extend along the first boundary where the front of the crack then meets the triangular region of softer phase. Crack blunting occurs in the critical temperature range to produce the observed enhancement of strength. The viscosity of glasses usually shows a pronounced temperature dependence and decreases typically by an order of magnitude for each temperature increase of 30–50 K. This explains the rapid fall-off in strength at higher temperatures as the softer phase becomes very fluid.

This type of behaviour has now been observed in several two-phase ceramic materials such as self-bonded silicon carbide, where plastic effects occur in the silicon phase, and in silicon nitride oxidised to form appreciable amounts of silica which again can deform plastically at particular temperatures (Davidge *et al.*, 1976). This raises the interesting technological possibility of designing multiphase ceramics (for duty at particular temperatures) which show a strength increase under stress rather than the usual decrease.

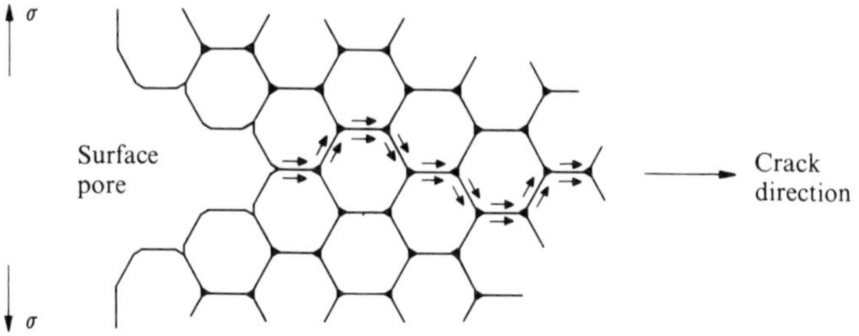

Fig. 6.20. Model for crack propagation at high temperatures in two-phase ceramics where a minor phase of lower softening temperature is situated at grain junctions. (After McLaren and Davidge, 1975.)

6.4 The development of strong ceramics

We noted in § 3.1 that a high theoretical strength is associated with a high density of strong bonds which is concomitant with a high stiffness and surface energy and a small lattice spacing. The same requirements for the strengths of practical materials are suggested from the Griffith equation, save that a small lattice spacing is replaced by a small flaw size. Additionally, fig. 1.13 indicates that a high flow stress is necessary if the high strength is to be coupled with good refractoriness.

There is no unique engineering ceramic in these respects and, apart from high strength, numerous other factors, such as cost, fabricability, chemical stability and thermal-shock resistance, need consideration for specific applications.

The desire for a high flow stress demands high purity in that impurities either as second phases or at grain boundaries tend to limit refractoriness. The intrinsic limitations on the flow stress favour materials with predominantly covalent bonding where there is a high Peierls–Nabarro stress for dislocation motion. Compounds including BN, SiC and Si_3N_4 must, therefore, have high potential as strong materials.

The above materials are also likely to meet the fracture strength criterion of a high stiffness, although little comment can be made at present concerning the factors controlling surface energy. The flaw size is best controlled by attention to microstructural parameters which are highly dependent on fabrication procedures. Here a minimum of porosity and a fine grain size are essential.

The ultimate practical strengths attainable in bulk engineering ceramics may be limited by the following typical parameters: $E=400$ GN m^{-2}, $\gamma_i=10$ J m^{-2}, $C=1$ μm. This suggests, from (3.23), a bend strength of ~ 2.2 GN m^{-2}. Values of about half this have already been realised in both hot-pressed silicon carbide and nitride. Small, rather than spectacular, improvements must, therefore, be predicted for the future.

7 Impact resistance and toughness

The discussion so far has been centred on strength. Most ceramic specimens or components fail eventually in a catastrophic manner although there are the obvious exceptions of particular fracture-mechanics specimens or localised cracking or chipping of components. It is interesting to consider what level of toughness, as say measured by the work of fracture, would produce a significant improvement in mechanical performance. Toughness can be considered at the microscopic or macroscopic levels. On a microscopic scale, for example under impact by small projectiles, any increase in toughness is useful in that the damaged area is reduced whenever the fracture strength of the material is exceeded locally. An increase in toughness is not necessarily useful on the macroscopic scale, however, unless the fracture energy is sufficiently large to prevent catastrophic failure of the component.

The argument can be quantified by considering a cube of material of dimension L stressed in tension to the fracture stress σ_f. The elastic energy in the cube is thus $\sigma_f^2 L^3/2E$. The energy to fracture the cube is that required to create two fracture surfaces of area L^2, that is $2L^2\gamma_f$. The critical condition to fracture the cube completely is therefore

$$L \geqslant \frac{4E\gamma_f}{\sigma_f^2}, \tag{7.1}$$

assuming that all the elastic energy is converted to surface energy. For typical values $E = 400$ GN m^{-2}, $\gamma_f = 20$ J m^{-2}, $\sigma_f = 400$ MN m^{-2}; $L \geqslant 200$ μm. Cubes larger than 200 μm could fracture catastrophically under these conditions. Noting that L is proportional to γ_f, a cube of 10 mm would require a work of fracture of 1 kJ m^{-2} to prevent catastrophic failure; this value is considerably greater than that observed for most ceramics.

In this chapter we discuss localised indentation and impact fracture, larger-scale impact behaviour and finally, possible ways of improving the toughness of ceramics.

7.1 Localised indentation and impact

The localised deformation and fracture that occurs under contact loading of ceramics is of considerable practical importance, ranging from

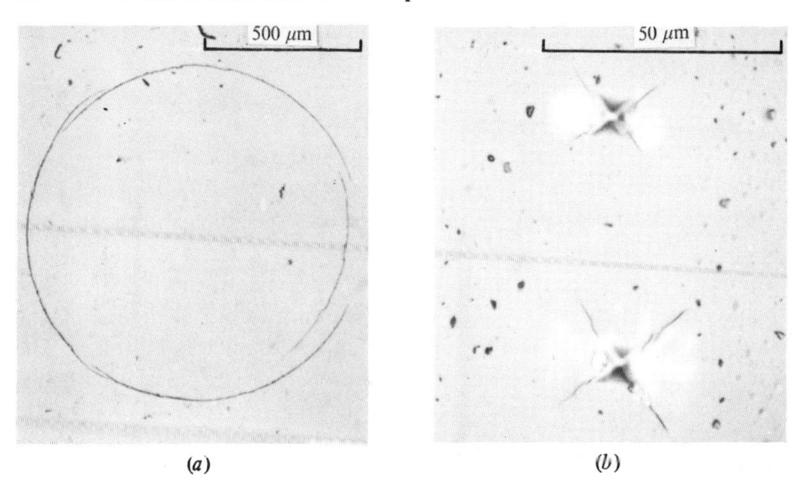

Fig. 7.1. Fracture induced by indentation. (*a*) Hertzian ring crack in WC–6% Co indented with steel sphere; (*b*) microcracking around diamond-pyramid indentations in glass.

damage by impact of small projectiles to the shaping of ceramics by grinding or drilling under point-loading conditions. Nevertheless, understanding of many practical problems is still at a developing stage even though the phenomenon of fracture under contact stresses is based on well-established principles. The first essential requirement is a detailed knowledge about the stress fields near the contact zone. These depend on the mechanical properties of the indenter and particularly on its shape. Although the stress fields are generally complex, significant tensile stresses occur for most cases of practical interest and fracture is thus common. We shall restrict the discussion to spherical and pointed indenters on bodies that are large compared to the indenter. Figure 7.1 shows how the mode of fracture depends on the indenter shape.

Calculation of the stresses set up during the contact of spherical bodies was due originally to Hertz (1881). For the geometry shown in fig. 7.2, the maximum tensile stress is on the surface at the boundary of the contact zone; this acts in a radial direction and is

$$\sigma_r = \frac{1-2\nu}{2}\frac{P}{\pi a^2}, \tag{7.2}$$

where P is the force between the sphere and plane surface and a the radius of the contact zone. P may be generated under static or dynamic loading conditions. The contact radius is given by

$$a^3 = \frac{4\kappa PR}{3E}, \tag{7.3}$$

where R is the sphere radius, and κ is a function of the elastic constants of the specimen (E, ν) and indenter (E', ν'):

$$\kappa=\frac{9}{16}\left[(1-\nu^2)+(1-\nu'^2)\frac{E}{E'}\right]. \tag{7.4}$$

When fracture occurs, its nature depends on several factors including the mechanical properties of the indenter and specimen and the size and shape of the indenter. The crack pattern is particularly sensitive to the absence or presence of plastic indentation beneath the indenter. The behaviour shown in fig. 7.2 relates to a condition of negligible plastic indentation and is typical for indenter sphere diameters 1–10 mm. The crack nucleates near the point of maximum tensile stress and usually just outside the circle of contact. It then propagates for a short distance normal to the surface before extending in diameter to form a cone crack. The crack path into the specimen follows closely the normal to the direction of the maximum principal tensile stress. The angle between the crack direction and the symmetry axis is $\sim 68^\circ$.

Experimentally it is observed that the critical force P_c to initiate a cone crack is proportional to the sphere diameter.

$$P_c=\beta_1 R \tag{7.5}$$

where β_1 is a constant depending on the surface energy (and in some cases on the original flaw size at the surface of the specimen). Consideration of (7.2) and (7.3), however, suggests that, for a criterion of critical stress, $P_c \propto R^2$. The explanation for this apparent anomaly lies in the fact that the crack may initiate, but then not necessarily propagate to form a cone crack unless the basic requirements of a Griffith energy-balance criterion are fulfilled (Frank and Lawn, 1967; Lawn and Wilshaw, 1975). A fracture-mechanics analysis indicates that the ring crack has to overcome an energy barrier before growing to a cone crack. Fracture initiation and significant propagation do not thus occur at the same

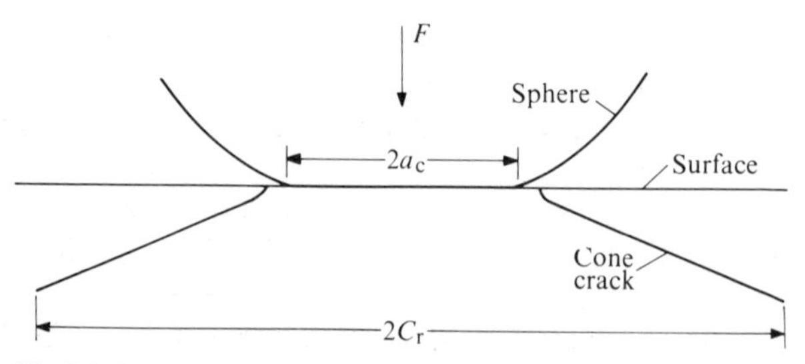

Fig. 7.2. Hertzian cone crack formed under spherical indenter.

applied force. The analysis also shows that the relation between the size of the cone crack and the applied force is

$$\gamma_p = \frac{\beta_2 P^2}{E C_r^3}, \tag{7.6}$$

where γ_p is the surface-energy requirement to propagate the crack and β_2 is a constant. It follows that γ_p may be estimated from experimental data relating P and C, and that once the cone crack is established the amount of damage sustained by the specimen depends particularly on the surface energy.

A different situation to that shown in fig. 7.2 obtains when the indenter radius is small and plastic indentation occurs. Following Evans and Wilshaw (1976), plastic indentation occurs when the maximum shear stress beneath the indenter exceeds a critical value. This stress is related to the maximum compressive stress σ_c equal to

$$\sigma_c = \frac{3P}{2\pi a^2}. \tag{7.7}$$

Eliminating a from (7.7) through (7.3), and equating σ_c to the hardness H gives the critical force for plastic indentation P_i as

$$P_i = \frac{\beta_3 \kappa^2 H^3 R^2}{E^2}. \tag{7.8}$$

The transition sphere radius R_c separating the two regions is obtained by equating (7.5) and (7.8):

$$R_c = \frac{\beta_1 E^2}{\beta_3 \kappa^2 H^3}. \tag{7.9}$$

It follows that R_c increases as the surface energy increases (through β_1) and as the hardness decreases. Typical values of R_c for glass are 10–100 μm.

A significant consequence for the cracking behaviour when $R < R_c$ is that the crack pattern for spherical indenters assumes many of the characteristics found under point indentation, fig. 7.1(*b*). The main features are that radial cracks now emanate from the contact zone, in addition to lateral cracks which lead to surface chipping as observed under high point-indentation forces. Analysis of this much more complex behaviour also permits estimates of surface energy to be made, and improvement in understanding here will lead to better control of important practical processes such as grinding.

7.2 Large-scale impact behaviour

Figure 7.3 illustrates test methods for determining the impact resistance of ceramic bars. In the Charpy-type test the moving pendulum strikes the specimen so that it deforms in a three-point bending mode. A variation is to use the specimen as the pendulum which strikes an anvil. In the Izod-type test the specimen is clamped at one end and deforms in a cantilever beam mode when struck by the pendulum.

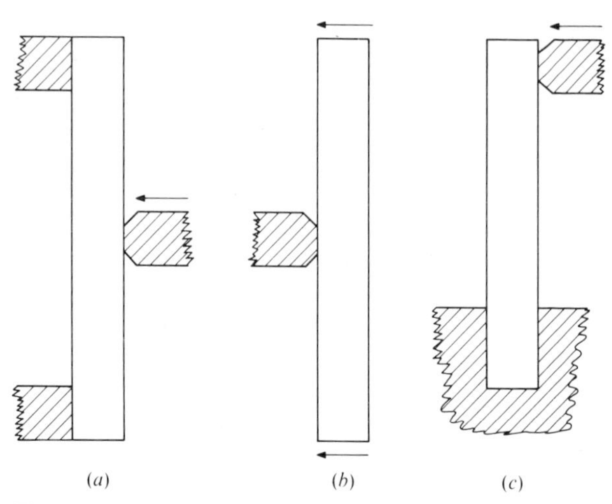

Fig. 7.3. Methods for determining the impact resistance of ceramics: (*a*) Charpy-type test (top view), the specimen is supported at its ends and struck at its centre by a pendulum pivoted above the specimen; (*b*) modified Charpy-type test (top view), the specimen acts as the pendulum and strikes a fixed anvil; (*c*) Izod-type test (side view), the specimen is clamped vertically and struck by a moving pendulum pivoted above the specimen. (After Davidge and Phillips, 1972.)

A common method of conducting Charpy or Izod tests on ceramics is to swing the pendulum from a fixed height and then measure the height reached after impact and fracture. The difference in potential energy U_p of the pendulum before and after impact is used to calculate the 'impact toughness' of the sample. When expressed in units of energy per unit area of fracture face produced $U_p/2A$, values in the range 1–10 kJ m^{-2} are obtained for typical ceramics. Fracture under these conditions is catastrophic; thus the impact toughness is not the same as the fracture surface energy, as is obvious from the large difference in numerical values between these quantities.

Two types of stress are set up in the specimen during impact: localised stresses around the point of impact and more macroscopic stresses as the specimen bends. Examination of broken specimens reveals that fracture is nucleated by the latter. During impact the specimen absorbs elastic energy until, at some location, the fracture stress under the test conditions is reached and the specimen breaks. For the Charpy specimen, fracture is initiated in the centre of the face opposite the impact region. The elastic energy U_e in the specimen at the point of fracture is

$$U_e = \frac{\sigma_f^2 Vg}{2E}. \tag{7.10}$$

V is the specimen volume and g a factor to account for the stress distribution in the specimen ($g = 1/9$ for square-section bars in three-point bending and 1/12 for round-section bars). Considering a typical specimen $10 \times 10 \times 100$ mm, with $\sigma_f = 400$ MN m^{-2}, and $E = 400$ GN m^{-2} gives $U_e = 0.22$ J from (7.10). Dividing this by the area of fracture face 2×10^{-4} m, indicates that the impact toughness is ~1 kJ m^{-2}. Additional energy may, however, be lost in the pendulum and specimen supports, localised deformation near the contact zone and in tossing the broken end(s) of the specimen (the toss factor). The recorded energy loss includes contributions from all of these sources.

An alternative test procedure is to increase the pendulum energy incrementally until the specimen fractures. This has the advantage that the incidental energy losses are minimised. Repeated blows may weaken the specimen but these effects are likely to be small (see § 9.2 for discussion of sub-critical crack growth). Incremental tests using the specimen as pendulum (fig. 7.3(*b*)) were conducted by Dinsdale *et al.* (1962) on a wide range of ceramic bars (77 mm long, ~8 mm diameter). Their results are expressed in fig. 7.4 in terms of the *measured* impact toughness $U_p/2A$, compared with a *calculated* impact toughness $U_e/2A$, from independent measurements of the material properties, using (7.10). The measured values are all greater than the calculated values, by ~50% on average. This difference is readily accounted for by noting that the calculated values incorporated data for the strength from *static*, three-point bending tests, which would be lower than the strength under dynamic conditions (see § 9.2); there will also be minor contributions from the other sources of energy loss mentioned above.

We conclude, therefore, that the impact tests described, if conducted in a controlled way, give an approximate measure of the dynamic strength, through (7.10), of brittle ceramics and not the fracture surface energy. The behaviour of ceramics under impact conditions is best improved by increasing strength rather than fracture energy.

However, as discussed above, an increase in work of fracture to ~1 kJ m^{-2} could prevent catastrophic failure under realistic conditions, and the conclusions reached here for normal brittle ceramics would not

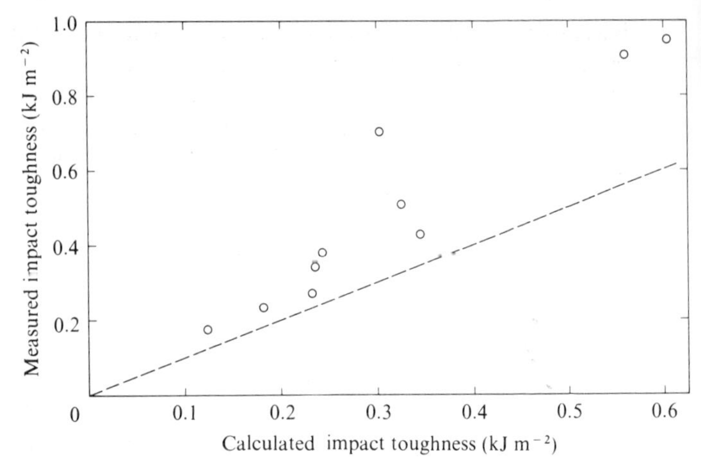

Fig. 7.4. Comparison of measured and calculated impact toughness for a range of ceramics, using apparatus of the type in fig. 7.3(*b*). (After Dinsdale *et al.*, 1962.)

necessarily hold. Some ways in which ceramists have sought to increase toughness are described next.

7.3 Tough ceramics

Inherent mechanisms

A major contributing factor to the surface energy of polycrystalline ceramics is the plastic deformation in regions adjacent to the fracture face, but a significant increase in toughness by this mechanism can be obtained only if there is a general plasticity over a larger region. This requires five independent slip systems, but there are no ceramics that meet this requirement at low temperatures. General plasticity may be achieved (table 4.2) in materials such as MgO and UO_2 at high temperatures where both primary and secondary slip occurs or in other ceramics with $\{111\}\langle 1\bar{1}0\rangle$ slip systems at high temperatures. Attempts have been made to modify the relative behaviour of the primary and secondary slip systems through the addition of particular additives. None of these, however, has met with success. It is possible, in principle, that the flow stress for primary and secondary slip could be made equal for particular ceramics but it seems unlikely that this could be applied generally.

Multiphase systems

One type of material where some success has been obtained in increasing the fracture energy is the ceramic–metal composite, or cermet, where there is a continuous metal phase surrounding the grains of ceramic. The best known example is the cemented carbide which comprises tungsten carbide grains surrounded by thin films of cobalt in amounts typically 5–15%. These find extensive application, for example, as tool-tip cutting materials. The important factor here is that during fracture the crack must pass through the metallic phase which, in deforming plastically, can absorb energy. The surface energy of these materials is typically 200–300 J m^{-2}, with a fracture strength >1 GN m^{-2}.

A recently discovered mechanism for increasing toughness has been discovered in zirconia which can exist in three structural forms: monoclinic, tetragonal and cubic. Garvie *et al.* (1975) prepared bodies of zirconia (containing calcia to stabilise the material in the cubic form) which were given a special treatment to induce additionally a fine dispersion of tetragonal second-phase particles <100 nm in size. Fracture energies <500 J m^{-2} are observed in these materials. It is believed that the main energy-absorbing mechanism is due to a martensitic, fast, diffusionless transformation as the crack passes near to the tetragonal zirconia particles which are metastable and convert to the monoclinic form. The existence of the metastable tetragonal phase is thought to be due in part to the lower surface energy of this phase compared with the monoclinic form and in part to the constraint of the matrix which opposes the formation of the less dense monoclinic phase. The transformation can thus occur locally with the absorption of energy whenever the surface constraints are removed, such as near to a fracture surface.

Fibre reinforced ceramics

The marked improvements in the mechanical properties of ceramics made possible by fibre reinforcement are indicated in fig. 7.5, which compares the stress/strain behaviour for glass and carbon reinforced glass (CRG) containing 40% of long aligned fibre. The stiffness of the material is increased by a factor of three because the fibres are much stiffer than the matrix. The strength is increased by a factor of seven because the fibres are much stronger than the matrix. The most spectacular improvement, however, is in the fracture energy of the material which is increased by about three orders of magnitude. This results from the controlled fracture behaviour, compared with the catastrophic failure of unreinforced glass. We discuss in more detail later the quantitative reasons for these properties, but fig. 7.6 illustrates a fracture face of CRG which shows clearly that many of the fibres have been pulled out from the opposite fracture face. In many ways the fracture process resembles that of wood and it is interesting to note that the fracture energy of wood is very similar to that of CRG (a few kJ m^{-2}).

There are a number of important requirements when fibres are to be incorporated into ceramic matrices. These include: (*a*) a suitable source of fibre at an economically attractive price; (*b*) fibres that generally are strong and stiff compared with the matrix phase; (*c*) an appropriate fabrication route which does not lead to degradation of the properties of the matrix or to damage of the fibres; (*d*) chemical compatibility between the fibres and matrix both during fabrication and in service; (*e*) physical compatibility between the fibre and matrix in terms, for example, of relative coefficients of thermal expansion; (*f*) an interface between the fibre and matrix that induces a fibrous type of fracture.

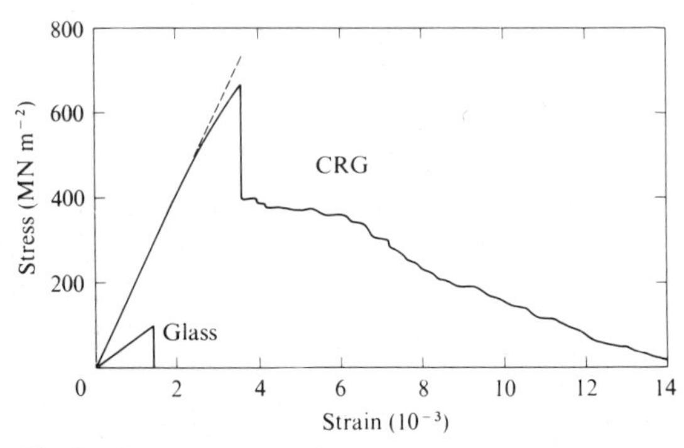

Fig. 7.5. Stress/strain data for glass and carbon fibre reinforced glass tested in bending. (After Bowen *et al.*, 1972.)

Several practical fibre reinforced ceramics exist including, with increasing temperature of application, asbestos fibre and glass fibre reinforced cement, carbon fibre reinforced glass and silicon carbide fibre reinforced silicon nitride, although only the cement-based materials are in current use. The geometrical arrangements of the fibres range from short fibres in random three-dimensional arrays to long fibres in unidirectional alignment. We consider mainly the latter where basic understanding is good.

It is instructive first to discuss the behaviour of an idealised system, fig. 7.7. It is assumed that the fibres are dispersed uniformly and unidirectionally in the matrix, and that a force is applied parallel to the fibres. For the moment it will be further assumed that there is no bond between the matrix and fibre so that the fibre can slip through the matrix at zero shear stress. Young's moduli and the volume fractions of the fibre and matrix are E_f, V_f and E_m, V_m. Application of the force results in an elastic deformation OA of the composite, as shown in the upper line,

Fig. 7.6. Scanning electron micrograph of fracture face of carbon fibre reinforced glass showing pull-out of fibres. (After Phillips, 1972.)

where the slope is proportional to Young's modulus for the composite E_c, and (cf. (2.13)).

$$E_c = E_f V_f + E_m V_m. \tag{7.11}$$

Equal strains are produced in the fibres and matrix. When the point A is reached on the stress/strain curve, corresponding to the failure strain of the matrix ε_{mu}, then all the load is transmitted to the fibres. This produces an extension at constant stress, A to B, where OB represents the elastic behaviour of the fibres alone. Further increase in stress produces deformation following BC until the failure strain of the fibres ε_{fu} is reached and ultimate failure occurs.

It should be noted that the mechanical behaviour of ceramic-based composites is different from composites with metallic or plastic matrices in that the failure strain of the matrix is usually much less than that of the fibres. This is a consequence of the lack of ductility in ceramics (compared with metals) and the small elastic strain at fracture (compared with plastics). In ceramic composites, therefore, the matrix fails before the fibres.

In real systems, either physical or chemical bonds exist between the fibres and the matrix. This means that part of the force is transmitted through the matrix even when the matrix is cracked. Consider the

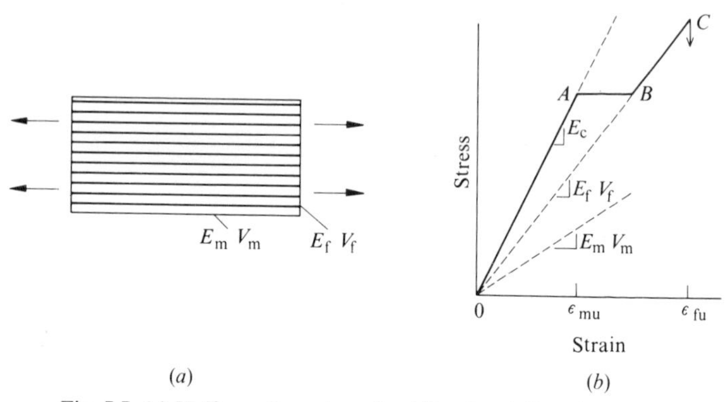

Fig. 7.7. (*a*) Uniform dispersion of unidirectional fibres in matrix; (*b*) stress/strain behaviour assuming no bonding between fibres and matrix and failure of matrix prior to that of fibres.

situation near a single crack in the matrix, fig. 7.8. Over a characteristic distance x from the fracture face the stress on the fibre will be greater than the average value and the stress in the matrix lower than the average value. The distance x depends on the shear strength of the interface τ. From a simple balance-of-forces argument it can be shown that x is given by

$$2\pi r_f \tau x = \pi r_f^2 \sigma_{mu} V_m / V_f \tag{7.12}$$

where r_f is the radius of the fibre and σ_{mu} is the ultimate strength of the matrix. This means that if the matrix has a well-defined breaking stress it will be broken up into a series of blocks of spacing between x and $2x$; this has been observed experimentally in a number of systems (Aveston *et al.*, 1975). In this situation the stress/strain curve (fig. 7.8) is similar to the previous curve except that the upper portion is shifted towards lower strains.

The analysis above assumes that the ultimate fracture stress of the matrix is unaffected by the presence of the fibres. In some cases, however, the presence of fibres results in a suppression of matrix cracking and the ultimate fracture stress of the matrix is then increased. Energy must be provided for a matrix crack to grow and this energy must be supplied from the elastic energy near to the tip of the crack plus any external work done on the specimen. The presence of fibres tends both to absorb energy as the crack grows and also to limit the supply of energy to a small region near to the growing crack. Cooper and Sillwood (1972)

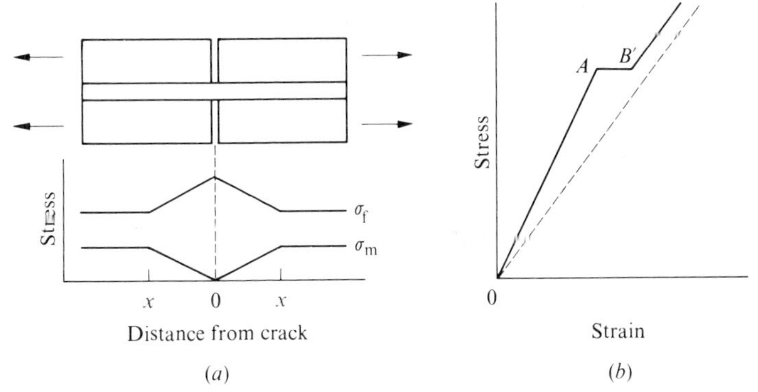

Fig. 7.8. (*a*) Stress distribution near a matrix crack for fibre bonded to matrix. (*b*) The corresponding stress/strain behaviour.

have considered the various factors contributing to the energy balance and calculated that the critical matrix strain is given by

$$\varepsilon_{mu} = \left(\frac{12\tau\gamma_{mf}E_f V_f^2}{E_c E_m^2 r_f V_m} \right)^{\frac{1}{3}} \tag{7.13}$$

where γ_{mf} is the fracture energy for the matrix. The suppression of matrix cracking can thus occur at high volume fractions of fibres of small radius. In glass containing 30% of carbon fibre by volume, for example, the presence of the fibres produces an increase in ε_{mu} of 2.5 times (Phillips *et al.*, 1972).

Table 7.1 gives illustrative data for the mechanical properties of carbon fibre reinforced glass and cement compared with the matrix phase. Agreement between experiment and theory for aligned composites is usually good. Figure 7.9 shows tensile stress/strain curves for both carbon and steel reinforced cement. The correlation between theory and experiment is remarkably good and the main divergence is that there is a more gradual change between the two major regions of the curve in the region of matrix cracking. This, however, is not surprising and simply reflects the fact that the matrix cracks over a range of stresses rather than at a unique value. The curves also show that the matrix-cracking stress increases with increasing fibre content.

Table 7.1. *Mechanical property data for unreinforced and reinforced ceramics (from Phillips* et al., *1972; Briggs* et al., *1974)*

	Young's modulus ($GN\ m^{-2}$)	Bend strength ($MN\ m^{-2}$)	Work of fracture ($kJ\ m^{-2}$)
Unreinforced Pyrex	60	100	0.004
Pyrex composite (50% carbon fibre by volume)	193	700	5
Unreinforced cement paste	13	10	0.02
Cement composite (9% carbon fibre by volume)	34	170	7.5

In fibre reinforced ceramics the work of fracture is high because pull-out of the fibres from the matrix absorbs large amounts of energy. For example, with the 40% by volume CRG shown in fig. 7.5 Phillips (1974) found $\gamma_f \sim 3\ kJ\ m^{-2}$. The total fracture process involves successively: (*a*) fracture of the matrix into blocks at the matrix-cracking stress; (*b*) fracture of the fibres near the ultimate fracture stress, with associated localised stress relaxation in the fibres; (*c*) pull-out of the fibres, and it is

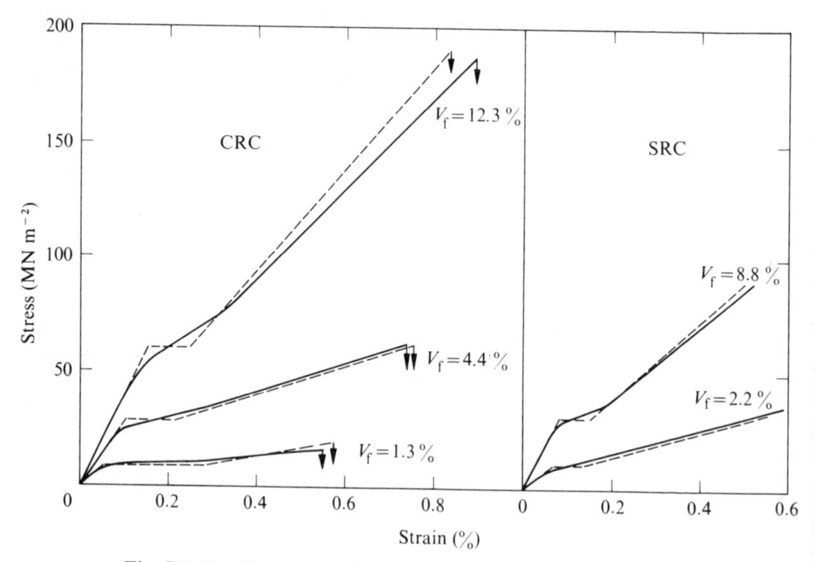

Fig. 7.9. Tensile stress/strain curves for cement reinforced with unidirectional carbon or steel fibres. Solid lines, experimental data; dashed lines, theory. (After Aveston *et al.*, 1975.)

possible to relate γ_f to these observed stages. Because of the stress-enhancement effect in the fibres, shown in fig. 7.8, the most likely origin of fibre fracture is near to a matrix crack. A significant proportion of fibres fail in this region and exhibit a short fibre pull-out length l_p, with a typical mean value for CRG $\sim 30\ \mu m$, increasing with decreasing strain rate. This compares with a matrix-cracking spacing of $\sim 90\ \mu m$. The dominant terms contributing to the work of fracture are the stress relaxation when the fibre breaks and the work to pull out the fractured fibres. The energy released when the stress in the fibres relaxes over a characteristic distance l_r is $\sigma_{fu}^2 V_f l_r / 2E_f$. The pull-out term is the product of the work to pull out a single fibre $\pi r l_p^2 \tau$, times the number of fibres per unit area $V_f/\pi r^2$. Thus

$$\gamma_f \sim \frac{\sigma_{fu}^2 V_f l_r}{2E_f} + \frac{\tau V_f l_p^2}{r}. \tag{7.14}$$

A high work of fracture is thus associated with a high volume fraction of fibres of small diameter, with the appropriate interfacial properties.

The extension of these theories to other cases of important practical significance, where the fibres may be randomly oriented throughout the matrix in short lengths and possibly in bundles, is a complex interpretative challenge and we mention only a few generalisations. The effect of having short rather than continuous fibres has little effect until the fibre length is reduced to about four times the matrix-cracking separation parameter x. For short fibres the strength is then limited not by the strength of the fibres but by the stress required to pull them out of the matrix. If the fibres are in random arrangements then clearly only a proportion of them can contribute to the property enhancements calculated for the case of aligned fibres. For example, the stiffness of a material with a random two-dimensional array of fibres is approximately one-third that indicated by (7.11).

8 Thermal stresses and fracture in ceramics

In most high-temperature engineering and refractory applications, ceramics are subjected to temperatures which vary throughout the component and which may change rapidly with time. In view of the importance of such applications it is essential to determine the magnitude of the stresses produced, whether these will lead to failure and, if so, the extent of the damage caused. We shall see that for engineering applications the conditions for fracture initiation are more important whereas for refractory ceramics the degree of damage is more significant.

The general thermal-shock behaviour of an engineering ceramic on rapid quenching is illustrated by the data in fig. 8.1. A series of 95%-alumina bars of 5 mm square cross-section and 20 mm long were heated to various temperatures < 1000 °C. Specimens were then quenched into water at 20 °C and the strength measured in three-point bending at 20 °C. The resulting data fall into a number of distinct regions. Unquenched specimens and those quenched from temperatures < 190 °C exhibit a mean strength of ~ 250 MN m^{-2}. Over a small temperature range 190–210 °C the fracture stress is either at the original level or at a much lower level, 100–150 MN m^{-2}. This suggests that once fracture is initiated a significant amount of damage is sustained by the specimen; this leads to a substantial drop in strength. For specimens quenched from temperatures of 200–300 °C the average strength remains approximately constant but quenching from higher temperatures results in a gradual diminution of strength.

We discuss below the origin and estimation of thermal stresses, thermal-shock parameters for the initiation of fracture and finally, a detailed explanation of the behaviour typified in fig. 8.1.

8.1 Thermal stresses

The principles involved in the calculation of thermal stresses are an extension of those already discussed in § 6.2 concerning the stresses developed in anisotropic or multi-phase ceramics on cooling after fabrication. Consider a rod of cylindrical section. If its temperature is raised uniformly from the ambient temperature T_0 to a higher temperature T then, provided that the rod is not constrained in any way, no stresses arise. The rod is free to expand or contract freely along its length.

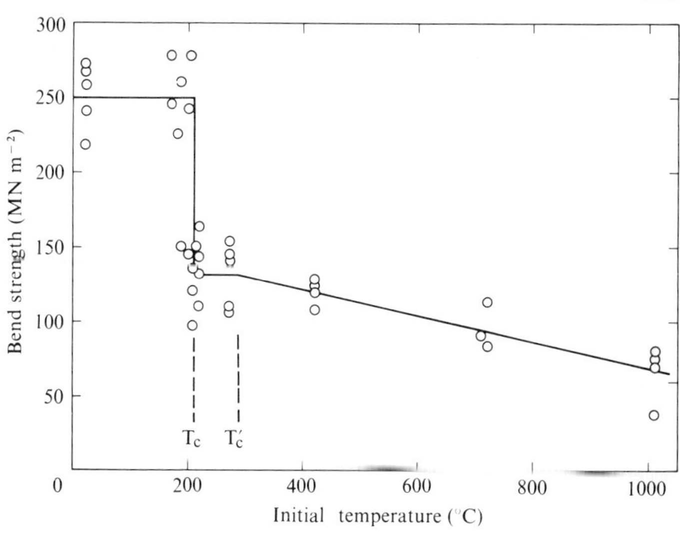

Fig. 8.1. Strength of 95% alumina bars after quenching in water at 20 °C from temperatures indicated. (After Davidge and Tappin, 1967.)

Similarly, if only the central portion of the rod is raised in temperature, as might be envisaged if the ceramic is being used as a furnace tube, then again, in the absence of constraints, no stresses occur in the direction parallel to the axis of the tube. If, on the other hand, the rod is rigidly constrained at its ends, as in fig. 8.2, whilst at temperature T_0, when it is heated to temperature T it attempts to expand, but under the influence of the constraints develops a compressive thermal stress σ_{ts} given by

$$\sigma_{ts} = E(T - T_0)\alpha. \qquad (8.1)$$

These stresses can be considerable even for small temperature differences. For alumina, using values of $E = 400$ GN m^{-2}, $\Delta T = T - T_0 = 100$ K and $\alpha = 9 \times 10^{-6}$ K^{-1} in (8.1) leads to a stress of 360 MN m^{-2}.

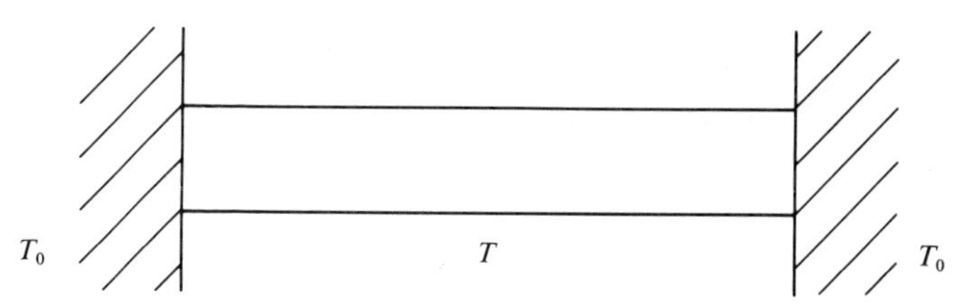

Fig. 8.2. Ceramic rod at temperature T rigidly restrained by supports at temperature T_0.

Also of practical significance are cases where thermal gradients are present throughout a component which is otherwise unconstrained. This, in general, leads to compressive and tensile stresses within the component, the magnitudes of which are determined by equilibrium considerations. (There are exceptions to this generalisation: for example, a thin slab with a linear thermal gradient across it can simply respond to the required expansions and contractions by uniform bending.) If a specimen is quenched from a temperature T in an environment at a lower temperature T_0 then, due to heat transfer from the body to the quenching medium, a tensile stress is set up in the surface of the component. This is because the surface attempts to contract, but cannot, due to the constraint from the bulk of the specimen. The maximum stress occurs under conditions where the temperature of the surface of the body attains that of the quenching medium almost instantaneously, whilst the conductivity of the material is such that an insignificant amount of heat has time to transfer from the interior to the surface. This *infinitely fast quench* is difficult to obtain in practice but might occur, for example, if a body of very low thermal conductivity is plunged from a high temperature into water. In this case the surface stress is given by

$$\sigma_{ts} = E\Delta T\alpha/(1-\nu), \tag{8.2}$$

where in comparison with (8.1) an extra term involving Poisson's ratio occurs associated with the biaxial nature of the stresses. This equation thus defines the maximum surface stress that can be produced by quenching.

In more practical examples one needs to consider two additional parameters: the heat-transfer coefficient between the material and the quenching medium h and the thermal conductivity of the material k. The theory of thermal stresses shows that the magnitude of the stresses throughout the material are usefully expressed in terms of a non-dimensional parameter referred to as Biot's modulus $\beta = ah/k$. It is convenient to consider idealised specimen geometries such as a plate of infinite extent or a cylindrical solid rod of infinite length; a is a specimen dimension such as the thickness of a slab or the radius of a cylinder. For the extreme case just considered, β is equal to infinity. Depending on the value of the Biot modulus the stress indicated in (8.2) is reduced proportionally by a factor ψ, such that

$$\sigma_{ts} = \psi E\Delta T\alpha/(1-\nu). \tag{8.3}$$

This factor has been calculated for a variety of geometries and fig. 8.3 shows curves (for various values of Biot modulus) which indicate how the stress-reduction factor at the surface of an infinitely long rod of circular section varies with time during quenching. It will be noted that the maximum stress occurs at a time which increases as the Biot modulus decreases; this time is zero for an infinitely fast quench. The analysis assumes that the material is isotropic, its behaviour is perfectly elastic and that the mechanical and physical properties and the heat-transfer coefficient are uniform at all temperatures.

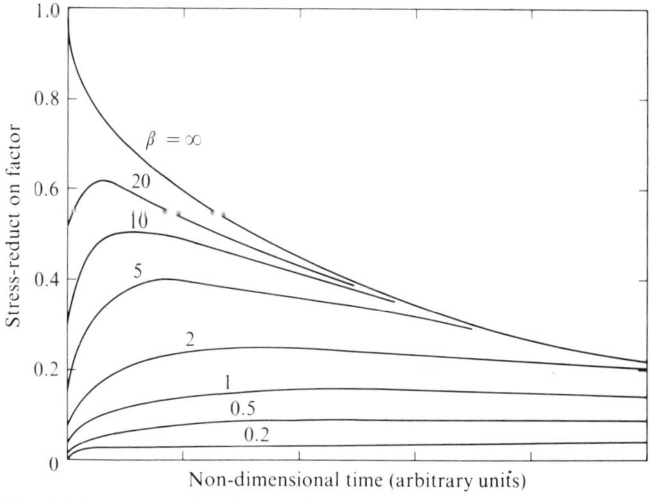

Fig. 8.3. Stress-reduction factor (ψ in (8.3)) versus non-dimensional time for various values of the Biot modulus. (After Jaeger, 1945.)

Under these conditions a parabolic temperature distribution is produced in a rod or plate. This is sketched for an infinitely long rod of circular cross-section, during cooling, in fig. 8.4, which shows that the maximum tensile stress is at the surface and the maximum compressive stress along the axis. For a section normal to the axis of the rod the compressive and tensile stresses must be in equilibrium. The stress at any position is proportional to the difference between the temperature at that position and the average temperature in the bar. By a simple balance-of-forces calculation it can be shown that in this case the maximum tensile and compressive stresses are equal and that the position of zero stress is situated at a distance $a/\sqrt{2}$ from the axis. (The equivalence of the maximum tensile and compressive stresses for the rod is purely coincidental; in the case of a plate the tensile stress is greater than the compressive stress.)

When the thermal stresses reach the strength of the material under the appropriate environmental conditions, fracture will be initiated. Due

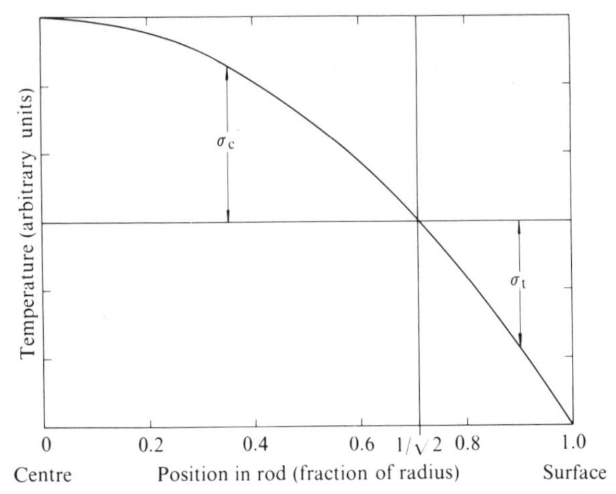

Fig. 8.4. Temperature and stress distribution across a section through an infinitely long, circular rod during cooling from its surface. (After Davidge and Tappin, 1967.)

account must of course be taken of the time under stress, size of specimen and the stress state in the specimen.

8.2 Thermal-shock parameters for initiation of fracture

In choosing between materials for a particular engineering application where thermal stresses are involved, it is of value to consider thermal-shock parameters that will give an indication of the fracture-initiation resistance of the competing materials. From the discussion so far it is clear that good thermal-shock resistance is associated with high values of strength and thermal conductivity and low values of Young's modulus and the coefficient of thermal expansion. The simplest parameter to consider is one relevant to very rapid quenches. This parameter R expresses the maximum temperature drop, given by (8.2), that a body can withstand in terms of the other variables in this equation. Table 8.1 gives data for a selection of engineering ceramics and includes values for R. There is a very wide range of values between the ceramics selected, from 625 K for hot-pressed Si_3N_4 to 40 K for hot-pressed BeO. In theory it should not be possible to break hot-pressed silicon nitride when quenched through a temperature interval of 625 K, however rapid the quench. The second thermal-shock parameter R' in table 8.1 is simply the product of the first parameter and the thermal conductivity of the material. This second parameter, although of greater complexity of

Table 8.1. *Mechanical and physical properties of some engineering ceramics and associated thermal-shock parameters*

Material	Bend strength σ ($MN\,m^{-2}$)	Young's modulus E ($GN\,m^{-2}$)	Poisson's ratio ν	Thermal expansion coefficient α, 0–1000 °C ($10^{-6}\,K^{-1}$)	Thermal conductivity k at 500 °C ($W\,m^{-1}\,K^{-1}$)	$R=\frac{\sigma(1-\nu)}{E\alpha}$ (K)	$R'=\frac{\sigma k(1-\nu)}{E\alpha}$ ($kW\,m^{-1}$)
Hot-pressed Si_3N_4	850	310	0.27	3.2	1[illegible]	625	11
Reaction-bonded Si_3N_4	240	220	0.27	3.2	15	250	3.7
Reaction-bonded SiC	500	410	0.24	4.3	84	215	18
Hot-pressed Al_2O_3	500	400	0.27	9.0	3	100	0.8
Hot-pressed BeO	200	400	0.34	8.5	63	40	2.4
Sintered WC (6% Co)	1400	600	0.26	4.9	85	350	30

physical significance than the first, is the more useful parameter of the two in that it relates to a more widespread range of practical conditions and does give a good order of merit of performance. The thermal-conductivity data in the table relate to 500 °C, but this parameter varies appreciably with temperature as shown in fig. 8.5. R' thus decreases, typically by a factor of four, from 0 to 1000 °C. Note that the relative order of merit indicated by the two thermal-shock parameters in table

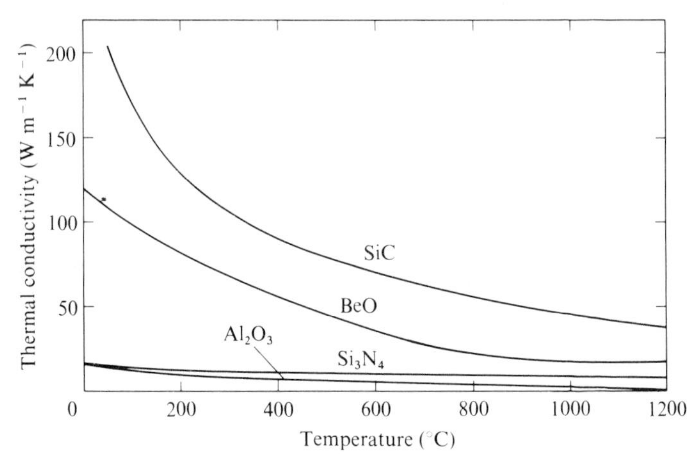

Fig. 8.5. Temperature dependence of the thermal conductivity of some engineering ceramics.

8.1 is very different. For example, reaction bonded SiC is three times inferior to hot-pressed Si_3N_4 in terms of R, but 50% superior in terms of R'. It is essential, therefore, to consider the precise conditions relating to the application when choosing between competing materials.

8.3 **Fracture initiation in alumina**

Provided that the necessary physical and mechanical parameters are known, it is possible, in principle, to calculate the maximum stresses that are expected for a particular component. This raises particular problems in that many of the relevant properties are likely to vary with temperature, in some cases significantly, so that the theory can only be used in the form discussed above to give approximate evaluations of the thermal stresses.

As an example we present some detailed thermal-shock data of Glenny and Royston (1958) for alumina; this is a 95% Al_2O_3 and is similar to that described in § 6.3, and in fig. 8.1. Specimens were thermally stressed under conditions of both rapid heating and rapid cooling, but we consider only the latter here. Cylindrical solid rods, in the diameter range 6–25 mm and 67 mm long, were fitted with loose end-caps so that on quenching the reduced heat flow through the ends of the cylinders would give stresses that approximated closely to those expected for infinitely long cylinders. The specimens were heated to a uniform temperature in a furnace and then rapidly plunged into a fluidised bed of silicon carbide powder maintained at 20 °C. Resulting values for β were <2. Specimens were examined for cracking after quenching with the aid of a dye penetrant. The approximate temperature interval required to fracture specimens of a given size was determined in a set of preliminary experiments. A batch of specimens of each size was then heated to a temperature just below that required to produce fracture and specimens were heated and quenched in a stepwise manner at 25 K intervals until all showed cracking. The mean temperature differential was computed for each batch of specimens. The scatter in the temperature differential values within a batch of specimens was generally $< \pm 30$ K.

Mean values for the critical temperature differential as a function of specimen diameter are indicated in table 8.2. This demonstrates clearly that the larger the diameter the more susceptible is the material to thermal-shock failure, as expected from fig. 8.3.

Table 8.2. *Variation in thermal-shock resistance of alumina rods as a function of diameter, Glenny and Royston (1958)*

Specimen diameter (mm)	6.4	9.5	12.7	19.1	25.4
Mean temperature drop for failure (K)	690	560	495	415	380

Glenny and Royston next computed the stresses at which failure had occurred. There were two complicating factors. The thermal conductivity decreases with increase in temperature by about a factor of two in the temperature range 20–800 °C, so that the Biot modulus varies in an inverse sense by the same factor. The coefficient of thermal expansion increases with temperature by a similar factor over this temperature range. These two effects both lead to higher thermal stresses, for an equivalent temperature interval, at the higher temperature. Unless, therefore, the temperature at which fracture occurs is known precisely the calculations can only be approximate. An iterative procedure was used to estimate the temperature T_i at which fracture was initiated and the mean values of the thermal conductivity and the thermal expansion coefficient used for the temperature interval $T-T_i$. The other relevant parameters such as Young's modulus, Poisson's ratio and the heat-transfer coefficient were sensibly constant over the conditions used in the experiments. Biot moduli of 0.92 for the 25.4 mm diameter bars and 0.30 for the 6.4 mm bars were obtained using these approximations, which from fig. 8.3 and (8.3) gave an estimate of the mean failure stress of ~ 170 MN m^{-2}. The thermal-failure stress increased slightly with increase in specimen size and this is contrary to expectations (to be discussed in § 9.1). It is possible that the insulating end-caps lead to proportionally larger heat losses for the large cylinders, causing a greater departure from theoretical expectations, and a corresponding overestimate of the strength of the larger specimens.

The strength of cylinders in bending as a function of temperature was also measured by Glenny and Royston. Little variation in strength with temperature was observed but the strength values did, in this case, decrease with increasing specimen size. The bend-strength values were approximately 20% greater than the computed thermal-failure stresses. However, the stresses during quenching are of a biaxial tensile nature whereas those in bending are uniaxial. We shall see from the data in § 9.5 that this is very close to the expected variation between the biaxial and unidirectional strength for this material. The agreement between experiment and theory can thus be considered very good when the complexity of the analysis and the simplifications used are borne in mind.

Much more sophisticated techniques are now available for estimating the thermal stresses in components of complex shape using finite-element methods of stress analysis and computer routines which allow the effects of variations in mechanical and physical properties to be estimated. This has resulted from the recent interest in using engineering ceramics as gas-turbine components.

8.4 Degree of damage

The next question to answer is how the observed degree of damage in a component, once fracture has been initiated, compares with that predicted theoretically. We return to the data in fig. 8.1. The amount and distribution of cracking in non-porous specimens can be revealed directly by a dye-penetrant technique. Figure 8.6 shows the crack patterns observed in 95% alumina specimens quenched from 220, 300 and 1000 °C. Two adjacent views are given for each quenching temperature: the external surface and an adjacent section through the mid-plane of the bar. The important points to note are that for temperatures just

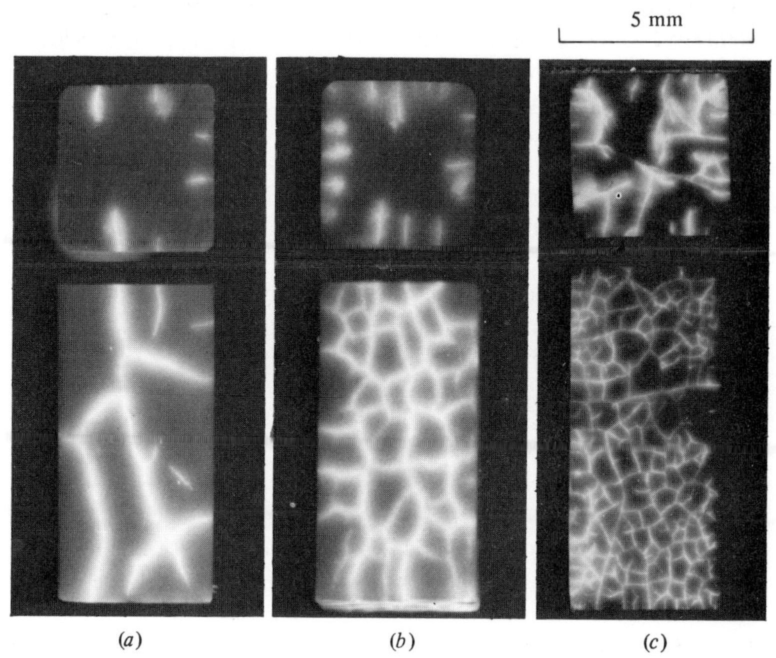

(*a*) (*b*) (*c*)

Fig. 8.6. Crack patterns, revealed by dye penetrant, in 95% alumina bars quenched in water at 20 °C from: (*a*) 220; (*b*) 300; (*c*) 1000 °C. Upper view: cross-section through centre of bar. Lower view: adjacent external surface. (After Davidge and Tappin, 1967.)

above T_c the amount of cracking is small but that the cracks penetrate roughly one-quarter of the distance through the bar, i.e. just past the boundary between the tensile and the compressive regions in the material (fig. 8.4). A comparison at this stage, when just a few crack sources operate, of the energy available for fracture with the energy required to produce the new surfaces shows that the former exceeds the latter several fold. As the quenching temperature is raised further above T_c an increasing number of cracks of similar depth of penetration are formed. This suggests strongly that fracture at these temperatures occurs by the successive operation of a number of flaws of decreasing initial severity. A comparison of the energies available and energies necessary for fracture for a range of aluminas quenched from 270 °C shows that the two energies are roughly in balance within a factor of two. Quenching from higher temperature produces a crazed surface network of cracks of similar size to that from 300 °C but the cracks now penetrate right to the centre of the specimen. This general type of behaviour has now been observed in a large number of engineering ceramics and appears to be the standard pattern. The observations are in qualitative support of the strength data, fig. 8.1, with regard to both the marked diminution of strength at T_c, the plateau region from T_c to T_c' and the subsequent fall at higher temperature.

In contrast, for much weaker refractory-type materials the thermal-shock behaviour in a similar test is very different. Figure 8.7 shows data for a 90% Al_2O_3 refractory with a strength about an order of magnitude weaker than the above alumina; specimens $114 \times 19 \times 19$ mm were

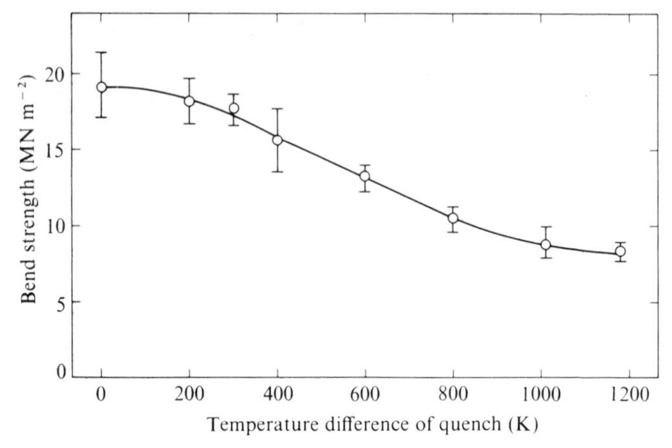

Fig. 8.7. Strength of 90% alumina refractory quenched in water at ~25 °C through temperature difference indicated. (After Larson *et al.*, 1974.)

quenched in water and the bend strength measured at 25 °C after drying. In this case there is a gradual fall-off in strength as the quenching temperature is raised suggesting that the cracks propagate in a more controlled manner.

The degree of damage sustained depends on both the magnitude and distribution of the stresses, and the fracture energy. The maximum amount of cracking is limited by the elastic strain energy available divided by the effective surface energy. This is related to earlier discussion for fracture under a uniform tensile stress (§ 3.6). For strong materials, a large amount of elastic energy is available at the fracture-initiation stress and the crack suffers a sudden extension with an attendant large decrease in strength of the material. On the other hand, for weak specimens or specimens which have already sustained some thermal-shock damage, the cracks will propagate in a more controlled manner in response to the variation of the strain with time. The first type of behaviour is typified by that shown for the dense alumina in fig. 8.1. at the lower temperatures, whilst the second type of behaviour is represented by the same alumina at higher temperatures, or the refractory alumina in fig. 8.7.

When a single crack source operates, the situation is similar to the case of unidirectional tension and we shall refer again to fig. 3.11. It is convenient to consider strain rather than stress. For a specimen following OK the critical strain to initiate fracture is OQ, and the crack propagates unstably to the point P. If such a cracked specimen is then subjected to a more severe thermal shock so that the strain is $>OQ$ or, alternatively, the strain is still increasing because the crack was nucleated at a strain lower than the maximum strain, the crack will not propagate further until a strain OR is exceeded. The size of the crack $2C_f$ at point P is estimated by equating the energies of the system at the initial and final stages. The relationship between stress and crack length is given by (Berry, 1960)

$$\sigma = \frac{AE\varepsilon}{(A+2\pi C^2)}. \tag{8.4}$$

Thus

$$\frac{\sigma_k{}^2}{2E}(A+2\pi C_0^2)+4\gamma_f C_0 = \frac{\sigma_l^2}{2E}(A+2\pi C_f^2)+4\gamma_f(C_f - C_0), \tag{8.5}$$

where σ_k and σ_l are the initial and final stresses. Note that the rather imprecise work of fracture has been used in this equation as an average value for the fracture energy. (For Al_2O_3, γ_f increases with increasing crack velocity.) Combining these equations to express C_f in terms of the failure strain ε_f and assuming that $C_f \gg C_0$ gives

$$C_f \sim \frac{AE\varepsilon_f^2}{8\gamma_f}. \tag{8.6}$$

A detailed description of this problem as relevant to thermal shock has been given by Hasselman (1969), although his model is subject to certain reservations. Hasselman assumed that the material contained a number of circular and uniformly distributed Griffith microcracks of identical dimensions and that the body was uniformly cooled with the external surfaces rigidly constrained to give a state of triaxial tensile stress. The model thus represents the worst possible case with all regions of the body under maximum stress. Crack propagation was assumed to occur by the simultaneous propagation of N cracks per unit volume, with negligible interactions between the stress fields of neighbouring cracks. Figure 8.8 expresses this idealised behaviour in terms of the relationships between crack length and the temperature difference and the resulting effect on strength. Comparison with the experimental data in fig. 8.1. shows excellent qualitative agreement.

An alternative fracture-mechanics approach, expressed in terms of K_I rather than γ_f, has been suggested by Evans (1975) in that crack arrest should occur when K_I at the crack tip reaches a critical value $K_{Ia} < K_{Ic}$. In certain crack-arrest situations relating to Hertzian and double-torsion tests $K_{Ia} \sim K_{Ic}$ so that, in terms of fig. 3.11, no overshoot should occur.

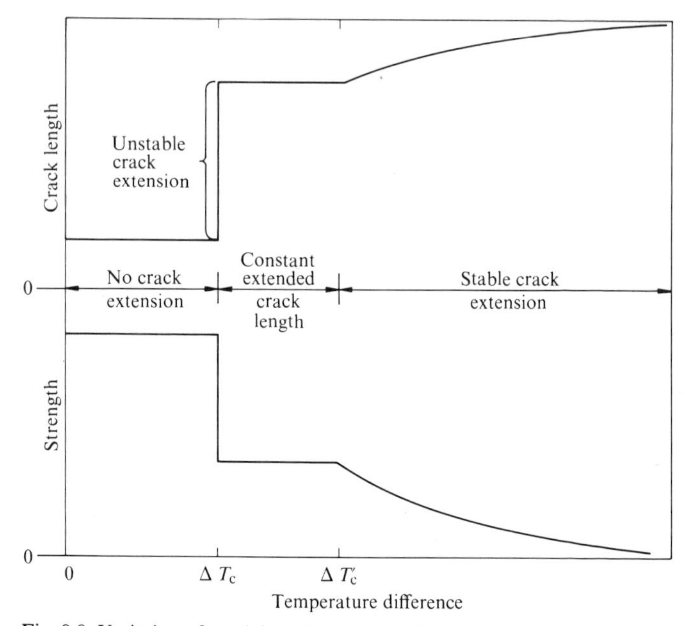

Fig. 8.8. Variation of crack length and strength for a strong ceramic subjected to thermal quenches of increasing severity. (After Hasselman 1969.)

However, there are no reliable thermal-shock data available on this point to distinguish between these models.

The above theories need modification to account for the actual stress distribution in quenched samples. Quenching produces both tensile and compressive stresses within the material and the crack, once nucleated, runs into a region of decreasing stress and then into a second region of increasing compressive stress. A more serious limitation, however, is that the distribution of crack sizes will mean that the largest cracks propagate first and, under conditions of multiple cracking, the crack pattern builds up through a succession of events rather than simultaneously. The distribution and severity of crack sources can at present only be revealed by a post-quenching examination and until further understanding is obtained advances in theory are limited. Other things being equal, the degree of damage (total amount of cracking) will be least in materials of high effective surface energy; the strength degradation on thermal shock, however, depends additionally on the number of active crack sources, and materials with a high density of uniform crack sources are desirable here.

9 Engineering design data

As the use of ceramics for engineering components becomes more widespread, it is clear that the materials scientist is to some extent failing in his duty to supply the design engineer with mechanical-property data in the most appropriate form. This leads to *ad hoc* and iterative design procedures. More rational techniques are now being developed based on improved understanding and a better presentation of data. The bulk of data for the strength of ceramics in the literature and technical information brochures relates to the maximum stress sustained in a bend test. This is a simple test to perform and is very useful for a rapid evaluation of materials on a comparative basis. Furthermore, basic theory has been developed to a stage where there is an excellent fundamental understanding of strength in terms of microstructural parameters, as already described.

Although these data are of use in choosing between possible materials they are of limited value to the design engineer. Additionally, information is required on (*a*) the statistical variation in strength or probability of failure at a particular stress level; (*b*) effects of long time loading under static or dynamic stresses; (*c*) effects of multiaxial stresses, all under the relevant conditions of environment and temperature.

The development of a ceramic component for a particular application involves a complex interaction between the materials and engineering aspects of the subject. Figure 9.1 gives a simple outline of the main processes. On the one hand there are the materials science aspects, where new materials are developed, their properties assessed and the material capability evaluated. On the other hand there are the engineering aspects, involving component design, stress analysis and the definition of the material requirements. Here we shall be involved mainly with materials science and will refer only briefly to the engineering side. It will be assumed, for instance, that the necessary stress analysis is feasible and can produce a satisfactory definition of the material requirements. Various methods of stress analysis are available, such as strain-gauge measurements, photoelastic observations or computer-aided theoretical analyses. In practice, however, the component may be too complex for these techniques and in this case an iterative design process may be necessary. Having defined both the material capability and material requirements, then a simple comparison should indicate whether the component satisfies the imposed criteria dictated by the operating conditions. Should the situation prove unsatisfactory there are two possibilities: either to develop or use an improved material, or to re-design the component such that the demands on the material are less stringent. These feedback processes are indicated in fig. 9.1.

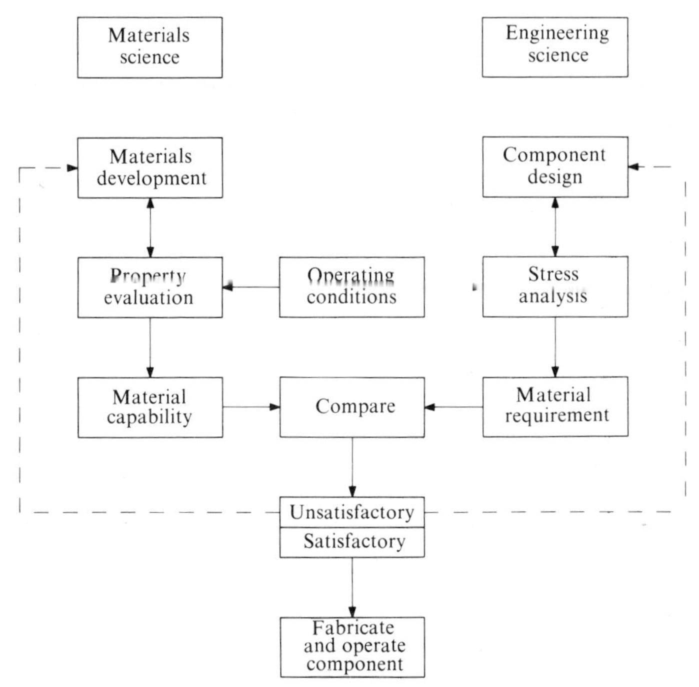

Fig. 9.1. The interaction between materials science and engineering science in the development of engineering ceramic components. (After Davidge, 1975.)

Much of the required information about the material capability can be presented on a strength/probability/time (SPT) diagram using data derived from simple experimental techniques. In this way, a component operating under known conditions of stress state, environment and temperature can be designed with regard to component lifetime and acceptable probability of failure. Considerations of statistical variations of strength and the variation of strength with time are given in the next sections.

9.1 Statistical variations in strength

Measurements of any physical parameter exhibit a spread of results when a series of tests is conducted under nominally similar conditions.

There are two basic reasons. First, there are inherent inaccuracies in the test method. In a mechanical property test these include the sensitivity of the load measuring device and the accuracy to which the dimensional measurements of the specimen can be made. Secondly, there is a genuine variation, from specimen to specimen, in the property being measured. In a well-conducted strength test on a single specimen, the value recorded should be accurate to within say $\pm 1\%$. However, on a small batch of ceramic specimens, strength variations of $\pm 25\%$ from the mean value are typically recorded. This should be compared with the spread of flow-stress values for a typical metal which is likely to be within just a few per cent. If ceramics are to be used for engineering applications, then clearly more detailed information on the statistical variations in strength is required. Hitherto, this information has generally not been required, mainly because there were few examples where ceramics were stressed to within a range near their ultimate strength.

No system can be made to operate with complete reliability and everyone accepts that cars, for example, are liable to mechanical failure. The degree of acceptance of failure, however, diminishes as the seriousness of the consequences increases; for systems such as aircraft or atomic reactors the probability of failure must be extremely low, typically less than from 1 in 10^4 to 1 in 10^6 per unit per year. On the other hand the failure of a component of a piece of machinery may be inconvenient rather than dangerous if the component is readily replaced, such as the exhaust system on a car. The main point is that for any component a degree of reliability can be demanded by the designer based on, for example, economic or safety arguments. It is the job of the component manufacturer to supply goods of the required reliability. Our aim in this section is to demonstrate how the ceramist can provide the necessary statistical information.

Ideally, one would like to understand strength variations in terms of the variables in the Griffith equation. For example, we might postulate that when the crack size is related to the grain size, the variation in strength should be related to the distribution in sizes of the largest grains. This could be so, but there is insufficient knowledge about grain-size distribution at extremely large grain sizes on which to base any calculations. Empirical approaches are, therefore, necessary. These are extremely powerful, however, and enable us to estimate survival probabilities at particular stress levels and to predict how strength varies with specimen size and stress distribution.

There are a very large number of statistical functions that can be applied to a set of data. Fortunately, when analysing the variation in the strength of ceramics, a particular function, due to the Swedish engineer

Weibull (1951), has been found to be applicable to many cases. Note that this is not universal and other statistical distributions may be appropriate to specific cases. The analysis below is thus presented as illustrative rather than of absolute applicability. The simplest form of Weibull's approach is based on a 'weakest link' model and is analogous to the breaking of a length of chain. Failure occurs when the weakest link breaks. In a series of chains of a particular length, the weakest link in each length is of different strength and this controls the variation in strength. The analogy can usefully be extended to the observation that there is an increase in strength with decreasing amount of material tested. This can be demonstrated simply be taking a long length of glass fibre and breaking it in tension. When the broken halves are re-tested they are stronger than the original length. Breaking the halves of the halves shows further increase in strength, and so on. This is clearly analogous to breaking a long chain into shorter and shorter pieces where the weakest link fails first, followed by the next strongest. Again one obtains increasing strength values for successive tests. In ceramics the links could thus represent small volumes of material containing a flaw and the weakest link is equivalent to the region with the largest flaw.

Suppose that under a stress σ there is a probability $P_s(l)$ that a length of chain l does not possess a link with a breaking stress $<\sigma$ and a corresponding probability $P_s(l')$ for a length l'. Then the probability $P_s(l+l')$ that a longer chain will survive under the stress σ is

$$P_s(l+l') = P_s(l)P_s(l'). \tag{9.1}$$

(For example if $P_s(l) = P_s(l') = \frac{1}{2}$, $P_s(l+l') = \frac{1}{4}$.) A parallel argument can be applied to survival of a volume V of ceramic, x times greater than a unit volume V_0.

$$P_s(V) = P_s(V_0)^x, \tag{9.2}$$

or

$$P_s(V) = \exp[x \ln P_s(V_0)]. \tag{9.3}$$

Weibull defined a risk of rupture R as

$$R = -x \ln P_s(V_0). \tag{9.4}$$

The risk of rupture for an infinitesimal volume $\mathrm{d}V$ is thus

$$\mathrm{d}R = f(\sigma)\mathrm{d}V, \tag{9.5}$$

since $\ln P_s(V_0)$ depends only on stress. Weibull postulated that

$$R = \left(\frac{\sigma - \sigma_u}{\sigma_0}\right)^m, \tag{9.6}$$

where σ_u, σ_0 and m can be regarded as material constants. (Although this expression appears cumbersome it is the simplest distribution function that satisfies the mathematical requirements. Of equal importance was Weibull's observation that it could be applied successfully to a wide range of problems, from the variations in the yield strength of steel, to the size of fly-ash particles and the stature of adult males born in the British Isles.) Thus

$$P_s(V) = \exp\left[-V\left(\frac{\sigma - \sigma_u}{\sigma_0}\right)^m\right]. \tag{9.7}$$

σ_u is the stress below which fracture is assumed to have zero probability, implying an upper limit to the flaw size (in many cases σ_u can conveniently be taken as zero); σ_0 is a normalising parameter of no physical significance; and m is a number, usually referred to as the Weibull modulus, which reflects the degree of variability in strength – the higher m is, the less variable is strength. Values for m of 5–20 are common for ceramics.

We emphasise again that this treatment is based partly on empirical reasoning although there is a small but increasing amount of theoretical justification for its use. Nevertheless it is widely and successfully used for ceramics as the examples following demonstrate. Time will either justify the Weibull approach on sound theoretical grounds and define its limitations, or supply a better alternative.

For the purpose of plotting data it is convenient to take logarithms of (9.7) twice and rearrange thus,

$$\ln\ln(1/P_s) = \ln V + m\ln(\sigma - \sigma_u) - m\ln\sigma_0. \tag{9.8}$$

Weibull-probability graph paper is available with $\ln\ln(1/P_s)$ as ordinate and $\ln\sigma$ as abscissa. As illustration, table 9.1 lists some tensile-strength data for a self-bonded silicon carbide. They have been ranked in ascending order. An important point to notice is that there is a spread within the sample of $\sim \pm 25\%$ about the mean. The mean strength is 309 MN m^{-2} and the standard deviation 40 MN m^{-2}. This degree of scatter is typical of engineering ceramics. In the analysis of such data through Weibull statistics, the plotting positions are obvious for the abscissa but not the ordinate. The ith ranked sample from a total of N can be plotted in several ways. The mean rank position is $P_s = 1 - i/(N+1)$; the median rank position is $P_s = 1 - (i - 0.3)/(N + 0.4)$. The reader is referred to Gumbell (1958) for the background to this terminology. For a sample of ten specimens the respective values of P_s are 0.91, 0.82, 0.73. . . and 0.93, 0.84, 0.74 . . . The differences are thus small but more significant at the extremes of the population where we have particular interest in making predictions about high probabilities of survival. Although the latter method is widely used, the former appears to be based on sounder statistical reasoning. The statistical arguments can be seen in the literature (Gumbel, 1958; Johnson, 1964). The data of table 9.1 are plotted as

Table 9.1. *Tensile-strength data for self-bonded silicon carbide*

Rank	1	2	3	4	5	6	7
Strength (MN m^{-2})	232	252	256	274	282	285	286
	8	9	10	11	12	13	14
	289	294	308	311	314	316	324
	15	16	17	18	19	20	21
	334	337	339	341	365	379	382

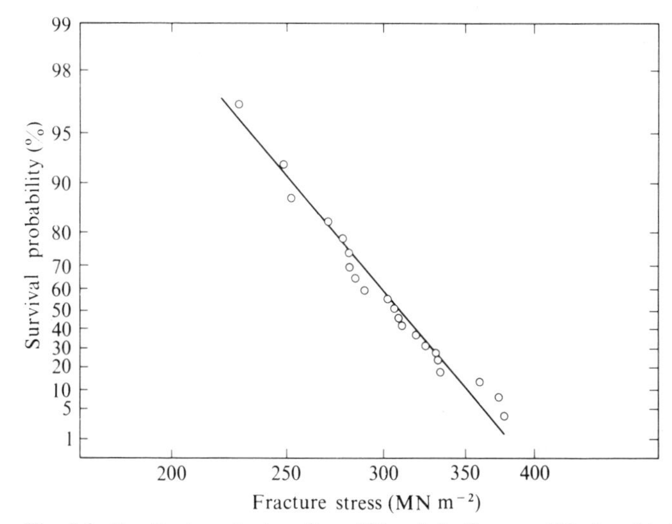

Fig. 9.2. Tensile-strength data for self-bonded silicon carbide in table 9.1 plotted according to (9.8) with $\sigma_u = 0$.

fig. 9.2, assuming $\sigma_u = 0$, using a least-squares analysis. The Weibull modulus m is given by the negative of the value of the slope and is 10 for this material. (It is necessary to take the negative of the slope in this particular plot because $\ln\ln(1/P_s)$ increases as P_s decreases.) Calculation of the Weibull modulus for a specific material and test condition thus enables predictions to be made about survival probabilities at particular stress levels. This is of value for engineering data but so far relates only to samples as tested experimentally. We extend below the arguments to cover the effects of specimen volume and the stress distribution in the specimen.

Under a uniform stress on two different specimens of volumes V_1, V_2 it follows from (9.7) that the stresses σ_{v1} σ_{v2} associated with the same probability of survival are given by $V_1\,\sigma_{v1}^m = V_2\,\sigma_{v2}^m$. (For simplicity in this and following calculations we assume $\sigma_u = 0$, i.e. there is no upper limit to the flaw size.) Thus

$$\frac{\sigma_{V_1}}{\sigma_{V_2}} = \left(\frac{V_2}{V_1}\right)^{1/m}, \tag{9.9}$$

which shows that the larger the specimen the lower its strength. As an example, the data in table 9.1 were obtained on specimens with a relatively small volume of material (115 mm³) under the maximum stress. The prediction from (9.9) with $m = 10$ is that the mean strength would be reduced from 309 MN m^{-2} to 254 MN m^{-2} for a tenfold increase in specimen volume. This clearly has important implications when making predictions for large components from data on small test pieces.

Equation (9.7) can be used also to account for stress variations throughout a component. It is necessary to know the variation of stress with volume and to evaluate the integral

$$P_s = \exp\left[-\int_V \left(\frac{\sigma}{\sigma_0}\right)^m \mathrm{d}V\right]. \tag{9.10}$$

Suppose we wish to compare the strength of a beam of material of breadth b, depth d and length l when tested in tension or three-point bending. In tension the integral is simply $bdl\,(\sigma_t/\sigma_0)^m$, where σ_t is the tensile strength. In bending, stress varies with distance from the point of maximum stress σ_{3b}, on the surface opposite the central knife-edge, both with distance into the beam and along the beam. Thus $\sigma = \sigma_{3b}(1 - 2d'/d)$ $(1 - 2l'/l)$ where d' is the distance from the surface into the beam and l' is the distance from the centre along the beam. At the neutral axis of the beam $\sigma = 0$ and that half of the specimen under a compressive stress can be neglected for the purpose of tensile failure. The integral in (9.10) thus reduces to

$$\left(\frac{\sigma_{3b}}{\sigma_0}\right)^m b\int_0^{d/2}(1-2d'/d)^m \mathrm{d}d' \cdot 2\int_0^{l/2}(1-2l'/l)^m \mathrm{d}l' = \left(\frac{\sigma_{3b}}{\sigma_0}\right)^m \frac{bdl}{2(m+1)^2}. \tag{9.11}$$

The ratio of the tensile and three-point bend strength for an equal probability of failure is thus

$$\frac{\sigma_{3b}}{\sigma_t}=[2(m+1)^2]^{1/m}. \qquad (9.12)$$

For $m=10$, $\sigma_{3b}/\sigma_t=1.73$ which shows why the bend strength of ceramics is considerably higher than the tensile strength.

The equivalent ratio for pure bending is easily shown to be

$$\frac{\sigma_{pb}}{\sigma_t}=[2(m+1)]^{1/m}. \qquad (9.13)$$

In four-point bending with the knife edges $l/4$ from the specimen ends, half of the beam is effectively in three-point bending and half in pure bending:

$$\frac{\sigma_{4b}}{\sigma_t}=\left[\frac{4(m+1)^2}{m+2}\right]^{1/m}. \qquad (9.14)$$

It is also possible to consider simultaneous variations in stress distribution and specimen volume through (9.10).

The discussion so far is based on the unjustified assumption that the flaws on the surface of the material have no greater effect than those in the interior. This is unlikely, recalling from chapter 3 that surface flaws are more serious than internal flaws of the same size, and from § 6.2 that the surface of a ceramic is more likely to contain large flaws from accidental damage than the interior. With a third factor, that the maximum stress is generally on the surface, it is often appropriate to ignore the interior of the material from the viewpoint of fracture initiation. In this case the volume term in the above equations is simply replaced by an area term. A useful discussion of this problem has been given by Davies (1973) for reaction-bonded silicon nitride.

9.2 Time dependence of strength

Time under stress is a very important parameter when considering the strength of ceramics. If a glass rod breaks under ambient conditions at a stress σ in a short time test, then an identical rod, when stressed to $\sim 0.75\sigma$ and the stress maintained, would fracture in a time about one hundred times longer than that taken in the original test. This sort of strength degradation with time is typical of oxide ceramics, and is clearly of great significance to engineering applications. As we shall see later, in carbide and nitride ceramics the time effects are much less severe at ambient temperature but at high temperatures can again be very significant. Let us consider first the basic theory behind the time dependence of strength; this can then be married with the statistical treatments described earlier.

Stress-intensity factor/crack velocity diagrams

The time-dependence degradation of strength is due to sub-critical crack growth occurring under stress, sometimes assisted by environmental factors. Of crucial importance is the relationship between stress-intensity factor and crack velocity. An idealised K/v diagram is shown in fig. 9.3. There are a number of features and different stages. There is sometimes a threshold value of K_I, K_{I_0}, below which no crack growth

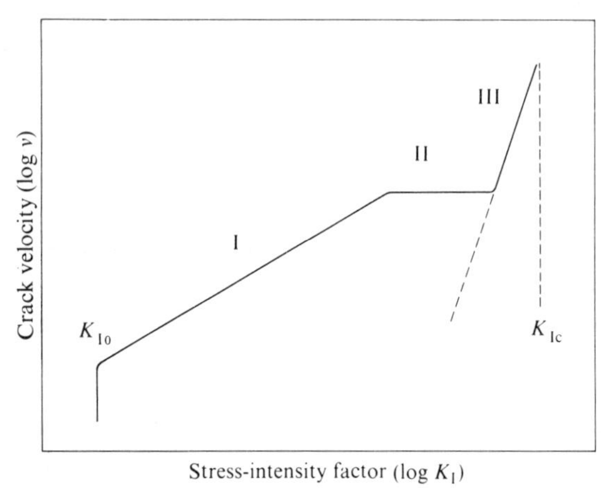

Fig. 9.3. Idealised stress-intensity factor/crack velocity diagram.

occurs. This has been observed particularly in glasses and there is indirect evidence of its occurrence in oxides. The existence of K_{I_0} can be justified on theoretical grounds. It is clearly advantageous to establish the existence or not of such a critical value in order to define a possible perfectly safe region of operation. This, however, is not easy because the crack velocities involved are particularly low. Regions I and II usually occur in oxides at ambient temperatures.

In region I:

$$v = \alpha_1 K_I^n,$$

whereas in region II: (9.15)

$$v = \alpha_2,$$

where the αs and n are constants. These effects have been identified with a stress-induced corrosion mechanism involving attack by water vapour. The rate of crack growth in region I is reaction-rate controlled.

In region II it depends on the diffusion of the corrosive species to the crack tip. Region III occurs at higher K_I values and is of rather academic interest in materials which exhibit the corrosive effects just mentioned. Region III is an intrinsic effect of the material and the slope of the K/v diagram here tends to be much steeper than in the corrosion-controlled region I. The origins of this stage are still being actively considered but possible mechanisms include dislocation-assisted slow crack growth, thermally activated crack growth and lattice trapping of the crack (somewhat related to the Peierls–Nabarro stress on dislocations). Further work is required to sort out the details. Finally, there is a critical K_I value, K_{Ic}, where fracture is virtually instantaneous. Definition of the K/v behaviour between the limits K_{I_0} and K_{Ic} enables estimates to be made of the time dependence of strength.

Curves of K/v may be generated by a number of the techniques described in chapter 3. The double-cantilever beam and the double-torsion tests are the most widely used. The latter is particularly useful in that data may be obtained in several ways. Most simply, the test machine can be operated at various strain rates and the force measured for each strain rate. Alternatively, the test machine can be held at constant deflection when the crack is moving at a particular velocity and the decay of force with time analysed, as described by Evans (1972), to give a large range of data on the K/v curve.

Some typical experimental K/v curves are shown in fig. 9.4. Oxides at ambient temperature exhibit the range of regions indicated in fig. 9.3 and the effect of variation in the concentration of water is clear. Wet environments lead to an extended region I and a region II at high velocities. Non-oxides generally show only a region III behaviour at low temperature and the effect of varying water concentration is usually negligible. Values for n in (9.15), the slope of the curve, are typically 10–20 for oxides in region I and ~ 100 for all materials in region III at ambient temperature. Region III behaviour dominates at high temperatures with n typically ~ 10.

Fundamental mechanisms

These data are convincing evidence that for oxides at ambient temperatures the rate of crack propagation is controlled by a chemical interaction between the ceramic and the water of the environment. The theory of Charles and Hillig (1962; Hillig and Charles, 1965) gives a good foundation for understanding this behaviour. This theory gives the velocity of chemically assisted crack growth:

$$v = v_0 \exp[-(E_0 + \gamma V_m/\rho - \sigma V^*)/kT]. \qquad (9.16)$$

Here E_0 is the activation energy of the reaction in the absence of stress. The second term represents the free-energy difference between a flat and curved surface (the tip of the crack) which tends to oppose crack motion: γ is the surface energy, V_m the molar volume of the solid and ρ the radius of curvature at the crack tip. The third term represents the stress-assisted effect where V^* is a constant defined as an activation volume, k is Boltzmann's constant and T is the absolute temperature.

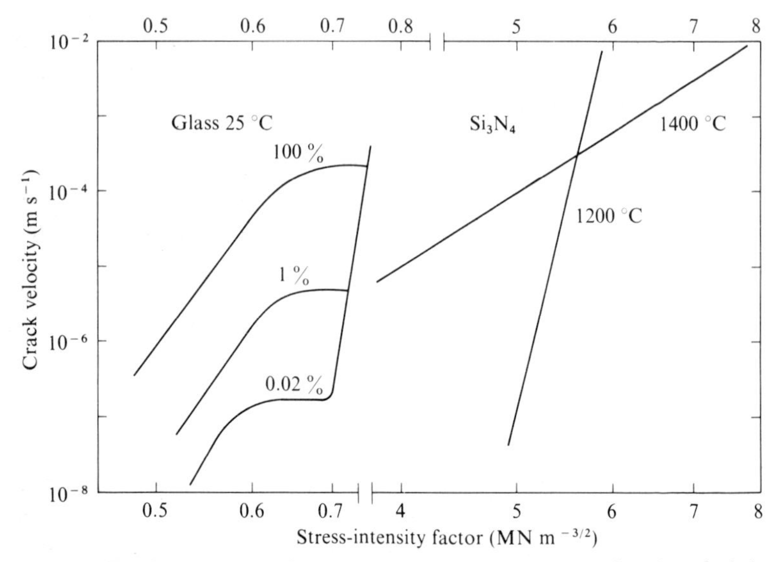

Fig. 9.4. K_I/v curves for glass at ambient temperature as a function of relative humidity (after Wiederhorn, 1967) and for hot-pressed silicon nitride at 1200 and 1400 °C (after Evans, 1974*b*.)

Equation (9.16) thus predicts a stress corrosion limit when the second and third terms are equal. Unfortunately this occurs at K values corresponding to extremely low crack velocities and is thus often difficult to determine experimentally. For glass, Charles and Hillig calculate $K_{I_0} = 0.17\ K_{Ic}$. The theory predicts an exponential relation between crack velocity and stress and the numerous experimental data now available fit this relation reasonably well. It is more customary in recent literature, however, to plot data as a power relationship as in (9.15) where the fit is slightly better. Although this discrepancy is not serious an improvement in understanding would be useful. The other aspects of the model such as the temperature dependence are also in good accordance with experimental data.

The Charles and Hillig model is based on the explicit assumption that the reaction rate between the corrosive species and the ceramic is controlled by a thermally activated process rather than the rate of supply of reactant to the crack tip. For dilute reactant concentrations or high crack velocities this assumption is not necessarily valid and here transport-controlled behaviour is expected. This should not be dependent on stress, again in agreement with the experimental data.

Unfortunately the success of the above model is not matched by those for region III behaviour. Several mechanisms have been proposed to explain region III effects, ranging from diffusional processes and plastic deformation to thermally activated crack growth based on a lattice trapping mechanism associated with the discrete nature of the lattice. It is still too early to present quantitative predictions or even to say which mechanisms are the more likely.

The important conclusions from an engineering design aspect are that for a wide range of ceramics both region I and region III behaviour are adequately represented by a simple power function. We shall thus proceed on this basis bearing in mind that significant advances in theoretical understanding are still required. Recent discussions are included in Wiederhorn *et al.* (1974), Lawn and Wilshaw (1975), and Evans and Langdon (1976).

Relationships between time to failure and stress

For the simple power relationship we need to discuss next the consequences for the delayed-fracture behaviour of ceramics. The simplest case to consider is that of the application of a constant stress. This induces a particular stress-intensity factor on a flaw, depending on its size, and the crack then grows at a velocity dictated by the K/v data. As the crack grows the stress-intensity factor increases and the crack velocity increases accordingly. Failure occurs when the stress-intensity factor on the most critical flaw reaches K_{Ic}, the critical stress-intensity factor.

Under constant σ the time to failure t is given by

$$t = \int_{C_i}^{C_c} \frac{dC}{v}. \qquad (9.17)$$

C_i is the initial crack size and C_c is a critical size such that $Y\sigma C_c^{\frac{1}{2}}$ equals K_{Ic}. From (3.17)

$$dC = \frac{2K_I}{\sigma^2 Y^2}\, dK_I, \qquad (9.18)$$

and substitution into (9.17) gives

$$t = \frac{2}{\sigma^2 Y^2} \int_{K_{Ii}}^{K_{Ic}} \frac{K_I}{v}\, dK_I. \qquad (9.19)$$

For materials exhibiting region I behaviour, recalling (9.15) and defining $K_{I\bullet}$ as the value separating regions I and II,

$$t=\frac{2}{\sigma^2 Y^2}\left(\frac{1}{\alpha_1}\int_{K_{Ii}}^{K_{I\bullet}} K_I^{(1-n)} dK_I+\frac{1}{\alpha_2}\int_{K_{I\bullet}}^{K_{Ic}} K_I \, dK_I\right). \quad (9.20)$$

A third term corresponding to region III may be added but this is generally negligible when regions I and II are present. Depending on the environment the second term (region II) is also often negligible and, in many ceramics tested under ambient conditions, the time to failure is controlled effectively by the behaviour in region I. In other materials region III dominates. Equation (9.20) reduces to

$$t=\frac{2}{\sigma^2 Y^2 \alpha_1 (n-2)}\left(K_{Ii}^{2-n}-K_{I\bullet}^{2-n}\right), \quad (9.21)$$

and, furthermore, because n is large (typically >10), $K_{Ii}^{2-n} \gg K_{I\bullet}^{2-n}$ and

$$t=\frac{2K_{Ii}^{2-n}}{\sigma^2 Y^2 \alpha_1 (n-2)}. \quad (9.22)$$

Substituting for K_{Ii} from (3.16) shows that for a particular specimen (C_i fixed) the product $t\sigma^n$ is constant and, thus, the ratio of the lifetimes $t_{\sigma i}$, $t_{\sigma j}$ at two sub-critical stresses σ_i, σ_j is given by

$$\left(\frac{\sigma_i}{\sigma_j}\right)^n=\frac{t_{\sigma_j}}{t_{\sigma_i}}. \quad (9.23)$$

This particularly simple relationship permits the straightforward calculation of a strength/probability/time diagram by combining the statistical and the time-dependent properties of ceramics.

9.3 Strength/probability/time (SPT) diagrams

Derivation of an SPT diagram requires data for the statistical variation of strength and a value for n. Strength variations are conveniently generated by breaking a number of samples at a specific strain rate, as in fig. 9.2. In this case it is necessary first to consider the connection between constant-strain-rate tests and delayed-fracture tests. Figure 9.5 presents idealised stress/time profiles for the two cases. The failure time under constant-strain-rate conditions is $t_{\dot{\varepsilon}}$. Had this same specimen been stressed instantaneously to σ_f it would have survived a much shorter lifetime t_σ, there being little sub-critical crack growth in the former case until stresses approach σ_f. It is readily shown that the ratio of these lifetimes is given by

$$t_{\dot{\varepsilon}}=(n+1)t_\sigma. \quad (9.24)$$

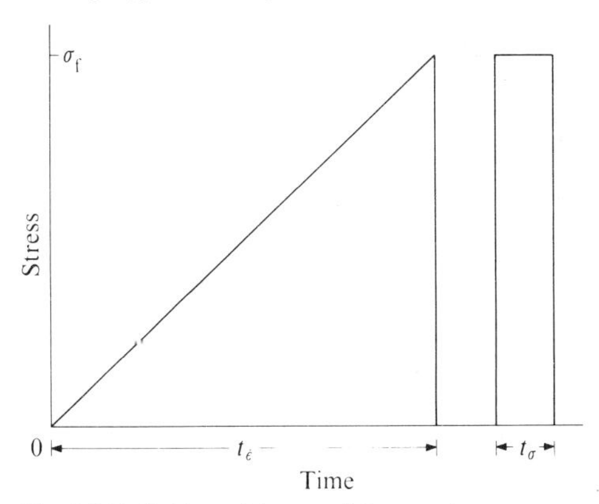

Fig. 9.5. Definition of times to failure under constant strain rate $t_{\dot{\varepsilon}}$ and constant stress t_σ.

A failure time under constant-stress conditions can thus be ascribed to each of the specimens fractured in constant-strain-rate tests. These times vary slightly because the weaker specimens fail in times shorter than those for the stronger specimens (by ~40% for the specimens in fig. 9.2). The failure stresses thus require normalisation to a constant failure time (say 1 s) through (9.23). The final step is then to construct failure lines for increasing times, again through (9.23). Each decade increase of time produces a failure line parallel to the initial line and, on Weibull-probability paper, these are equispaced. The spacing increases as n decreases. With $n = 10$ a reduction in stress of 21% is obtained for each decade

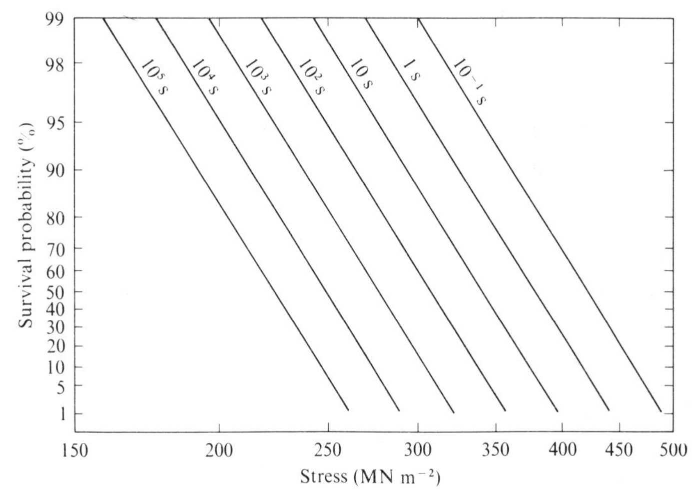

Fig. 9.6. An SPT diagram for 95% alumina tested in bending at 20 °C. (After Davidge *et al.*, 1973.)

increase in time, and for $n = 100$ a reduction of 2.3%. An SPT diagram for alumina under ambient conditions is shown in fig. 9.6.

It should be emphasised that the diagram is valid only for the conditions under which the specimens are tested including specimen size, mode of stressing and ambient conditions. Allowance for specimen size can, however, be made by the analyses in § 9.1. The predictive use of the diagram is evident from a simple example. For engineering requirements

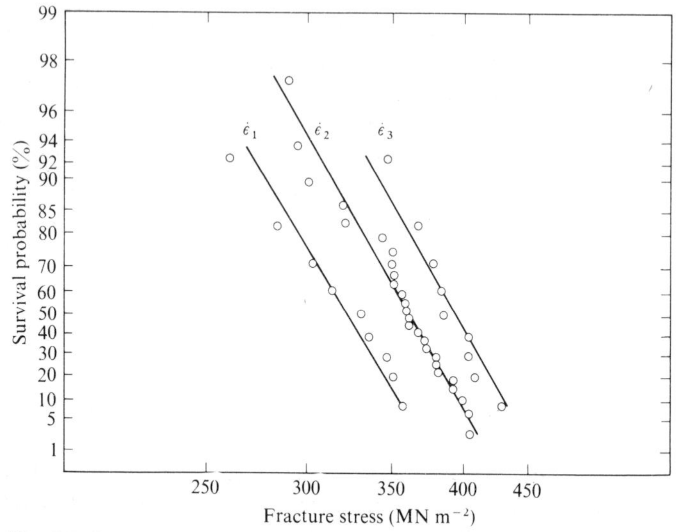

Fig. 9.7. Strain-rate dependence of strength for 95% alumina at 20 °C: $\dot{\varepsilon}_1$, 2×10^{-6} s^{-1}; $\dot{\varepsilon}_2$, 2×10^{-5} s^{-1}; $\dot{\varepsilon}_3$, 2×10^{-4} s^{-1}. (After Davidge *et al.*, 1973.)

of 99% survival probability and a component lifetime of 10^5 s one can estimate the permissible working stress as 160 MN m^{-2}. (This compares with a mean short-term strength of 380 MN m^{-2}.) If the stress level on the component is less than this then the situation is satisfactory. If not, then either further materials development or engineering design is required as indicated by the two options in fig. 9.1.

So far we have assumed that n is derived directly from K/v diagrams. However, once the general form of the K/v diagram has been established for a material it is often more convenient to calculate n from other test data such as the strain-rate dependence of strength. Figure 9.7 gives data for the alumina used above. For specimens failing at $\sigma_{\dot{\varepsilon}_i}$, $\sigma_{\dot{\varepsilon}_j}$ in times $t_{\dot{\varepsilon}_i}$, $t_{\dot{\varepsilon}_j}$ when tested at $\dot{\varepsilon}_i$, $\dot{\varepsilon}_j$,

$$\sigma_{\dot{\varepsilon}_i} = E\dot{\varepsilon}_i t_{\dot{\varepsilon}_i}$$
$$\sigma_{\dot{\varepsilon}_j} = E\dot{\varepsilon}_j t_{\dot{\varepsilon}_j}. \qquad (9.25)$$

Eliminating time through (9.24) and (9.23) gives

$$\left(\frac{\sigma_{\dot{\varepsilon}_i}}{\sigma_{\dot{\varepsilon}_j}}\right)^{n+1} = \frac{\dot{\varepsilon}_i}{\dot{\varepsilon}_j}, \tag{9.26}$$

and hence an independent estimate of n. Note that both the slopes and spacings of the lines in fig. 9.7 are slightly different from those in the SPT diagram, but the above mathematical relationships can be used to check that the data are self-consistent.

A particularly worrying aspect of the properties of ceramics, with reference to engineering applications, is that there appears to be no absolutely safe working stress, so that there is always a small but finite probability of failure at any combination of stress and time. Proof-testing is thus an attractive concept to investigate with the hope that the weaker components from a batch may be eliminated, with a guarantee on the performance of the survivors. Let us see whether this hope can be realised for engineering ceramics.

9.4 Proof-testing

Proof-testing is a technique whereby individual samples from a batch, that are not up to specification can be removed before entering service. In its simplest form a proof test would subject components to particular conditions relating to performance requirements, whereupon some would fail, with the remainder expected to perform satisfactorily in service. This would be acceptable for ceramics only if there were no time dependence of strength. The illustrative data in fig. 9.8 reveal some of the problems. Two sets of specimens were broken. The first were fractured under a constant strain rate (mean failure time 85 s) and the data gave the upper single line plotted on Weibull graph paper. The second set were fractured under proof-test conditions. The proof stress σ_p was selected at the 70% survival level and specimens were raised to this stress at the original strain rate, below which the expected 30% of specimens broke. The survivors were then held at σ_p for times < 10 s during which time another 40% of specimens broke. The second survivors were finally fractured by a further increase in stress at the original strain rate. Two points should be noted. First, the fraction of specimens failing at σ_p depends on the time chosen for the proof test. Secondly, the proof test weakens markedly the weaker of the survivors of the proof test. Both these effects are predicted by the theory described above and quantitative agreement between the two sets of data is obtained using experimentally determined values of m and n.

During a realistic proof test only a small fraction of the specimens would be rejected through an appropriate selection of proof-test conditions. In spite of the weakening effect of the proof test on the weaker survivors, the strength distribution of the survivors is greatly improved compared with that of the original population. Furthermore, it is now possible under *ideal* conditions to give a *guarantee* for the performance

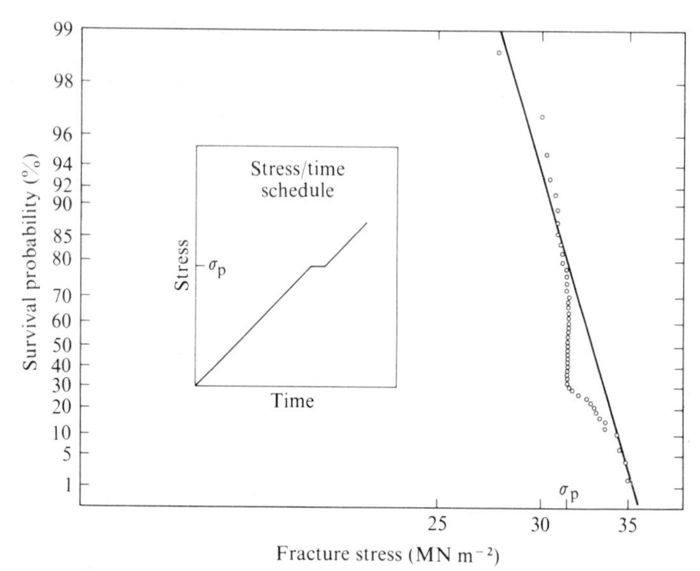

Fig. 9.8. Effects of proof-testing on the strength of a porous vitreous-bonded alumina ceramic at 20 °C. The solid line corresponds to specimens tested at a constant strain rate. The experimental points are for specimens subjected to the stress/time schedule indicated.

of all the survivors. The stress-intensity factor K_{Ip} on the surviving specimens for an *instantaneous* removal of the proof stress, is given by

$$K_{Ip} = Y\sigma_p C_p^{\frac{1}{2}} < K_{Ic}, \tag{9.27}$$

where C_p is the crack size after proof-testing (the crack size may have increased during the test). Note that $K_{Ip} < K_{Ic}$ otherwise failure would have occurred. In subsequent service under an applied stress $\sigma_a (< \sigma_p)$ the corresponding stress-intensity factor value K_{Ia} is

$$K_{Ia} = Y\sigma_a C_p^{\frac{1}{2}}. \tag{9.28}$$

Thus

$$K_{Ia} \leqslant \frac{\sigma_a}{\sigma_p} K_{Ic}. \tag{9.29}$$

The minimum time to failure t_{min} at σ_a can be estimated by substituting the maximum value of K_{Ia} from (9.29) into (9.22):

$$t_{min} = \frac{2K_{Ic}^{2-n}(\sigma_a/\sigma_p)^{2-n}}{\sigma_a^2 Y^2 \alpha_1 (n-2)}. \tag{9.30}$$

The practical significance of this equation is best appreciated by noting that for a given ratio of the proof stress to the required working stress

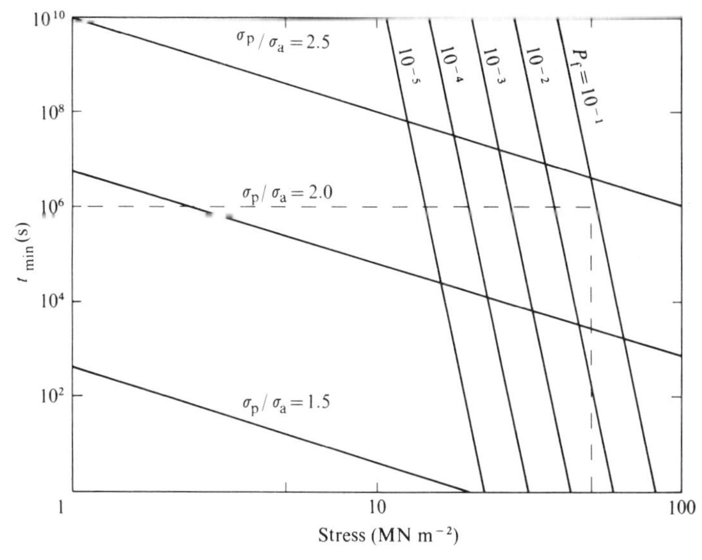

Fig. 9.9. Proof-test failure diagram for glass at ambient temperature. (After Wiederhorn, 1974.)

(proof-test ratio) the minimum time to failure is proportional to σ_a^{-2}. Plots of t_{min} v. σ_a thus give a series of straight lines, on logarithmic paper, of slope -2. An example for glass is shown in fig. 9.9. Superimposed are failure-probability data obtained from independent measurements. Suppose for example it is required to operate a component for 10^6 s at a stress of 50 MN m^{-2}. Application of a proof-test ratio of 2.4 ($\sigma_p = 120$ MN m^{-2}) would give the required performance of the survivors, $\sim 9\%$ of the specimens having failed the proof-test.

Ideally proof-testing should be conducted under conditions where sub-critical crack growth is absent. The effects of environment could be eliminated by testing in vacuum or at very low temperatures but this would still leave the intrinsic crack growth mechanisms (region III) which are however, less significant. When crack growth occurs during the proof test, growth at the proof stress is accounted for automatically by the above analysis. Unfortunately any crack growth on *unloading* from the proof stress is not accounted for and, although this effect may be small, it effectively nullifies the guarantee of performance, which can now only be expressed in statistical terms.

9.5 Strength under multiaxial stresses

Most experimental and theoretical data for the strength of ceramics relate to a simple condition of uniaxial tensile stress. Many engineering ceramic components, however, are not subjected to this simple stress state and a multiaxial-stress state is much more common. It is thus essential to have a good appraisal and understanding of the effects of multiaxial stresses. Unfortunately, this is a confusing field for the non-specialist in that there is a wealth of theories and hypotheses which far outweigh the dearth of reliable experimental data. The biaxial-stress state is the simplest and of the most general relevance. We shall restrict our discussion to the tensile/tensile and tensile/compressive states and consider their effect on the propagation condition for sharp flaws. As yet there is no satisfactory treatment for the general state of stress shown in fig. 2.1.

It is important first to distinguish between two situations: the effect of biaxial stresses on a specimen containing a single flaw and that on specimens containing the common random distribution of flaws. The latter is the case of greater practical significance but understanding here is dependent on that for the former case.

Specimens with single flaws

In Griffith's original treatment (discussed in § 3.2) which was based on an energy-balance criterion, consideration was given to a thin sheet of material containing a single narrow crack under a uniform biaxial tensile stress. Griffith concluded that rupture is determined entirely by the stress perpendicular to the crack, so that the uniaxial and biaxial tensile strengths should be equal. In his second paper Griffith admitted that the solution to the calculation of the strain energy presented in his original paper was incorrect because the expressions used for the stresses produced values at infinity that differed from the proposed uniform stress at infinity. Although these differences were infinitesimal they made a finite contribution to the energy on integration around the infinite boundary. Unfortunately, Griffith did not give sufficient details of the calculation for his reasoning to be followed.

Griffith's second paper was devoted to a discussion of the strength of brittle materials under various types of applied stress ranging from simple tension and compression through intermediate biaxial-stress states. In sympathy with his original theory Griffith made the hypothesis that a general condition for rupture should be the attainment of a specific tensile stress at the tip of the crack. If the strength under a specific stress condition were thus known then the strength under the

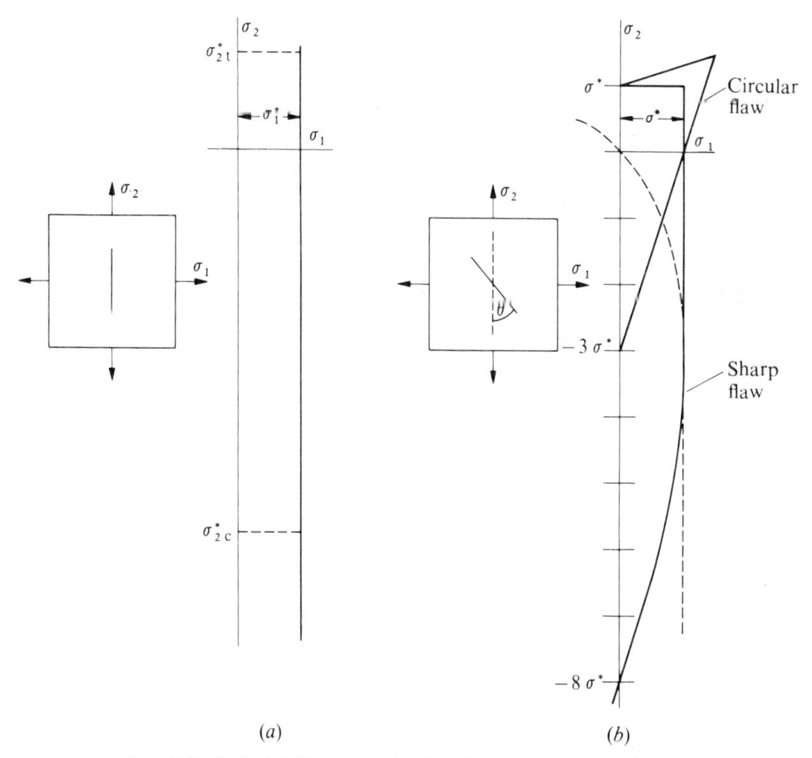

Fig. 9.10. Griffith failure criterion for sharp flaws under biaxial stress when (*a*) crack lies normal to σ_1 direction and (*b*) crack orientation is allowed to vary. Also shown in (*b*) is the case for a circular flaw.

other conditions could be estimated. This discussion is thus based on a maximum-tensile-stress criterion and not an energy-balance method as in his original paper.

Two cases considered by Griffith for sharp cracks are illustrated in fig. 9.10. In the first case the crack is situated normal to the stress σ_1, and parallel to the stress σ_2. Failure occurs when σ_1 reaches a critical value σ_1^*, dictated by the parameters in (3.11). In the ideal situation σ_2 has no influence, although in practice at high positive or negative values of σ_2 failure occurs for other reasons. Under tension failure occurs at σ_{2t}^* due to the propagation of fracture from flaws other than the one considered.

Under high compressive stresses failure occurs at σ^*_{2c} because of parasitic stresses introduced by the contact surfaces between the specimen and the compression platens.

In the second case the orientation of the flaw is allowed to vary and the failure criterion estimated for the orientation of the flaw that gives the maximum stress concentration. There are now several regimes of failure depending on the ratio of σ_2/σ_1. (σ_1 is assumed >0.) When $1 \geqslant \sigma_2/\sigma_1 \geqslant -3$ the failure criterion is identical to the above with $\sigma_1 = \sigma^*$, the tensile strength, irrespective of the value of σ_2; $\theta = 0°$, as previously. When $\sigma_2/\sigma_1 > 1$ failure occurs with $\theta = 90°$ at $\sigma_2 = \sigma^*$. When $\sigma_2/\sigma_1 < -3$ failure occurs at $\sigma_1 < \sigma^*$. The failure criterion is now

$$(\sigma_1 - \sigma_2)^2 + 8\sigma^*(\sigma_1 + \sigma_2) = 0. \tag{9.31}$$

This section of the curve is parabolic, and θ is now not zero but given by

$$\cos 2\theta = -(\sigma_1 - \sigma_2)/2(\sigma_1 + \sigma_2). \tag{9.32}$$

Thus θ varies from 0° when $\sigma_2/\sigma_1 = -3$, to 30° for uniaxial compression.

This critical-tensile-stress criterion has since been extended to the case of an elliptical flaw (Babel and Sines, 1968). The sharp flaw is one extreme where the ellipse is very narrow. At the other extreme is the circular flaw. The general tendency on going from a flat flaw to a circular flaw is to reduce the stresses required to attain the critical stress under tensile/compressive stresses and to increase them under tensile/tensile stresses. Calculations for the circular flaws are shown in fig. 9.10 and the results for intermediate flaw shapes lie between these extremes.

The basic premise of Griffith that a stress parallel to the crack does not affect the fracture strength measured normal to the crack has been disputed in the literature. Swedlow (1965), for example, develops a failure criterion based on strain-energy considerations, as in Griffith's original treatment, and claims to demonstrate that in a biaxial-stress state the stress applied parallel to the crack *does* have an effect on the fracture stress. According to Swedlow's calculations the tensile/tensile strength differs from that predicted by Griffith depending on the value of Poisson's ratio. For $\nu < 0.25$, the biaxial strength is greater than the Griffith criterion and for $\nu > 0.25$, less than the Griffith criterion.

There is still no general agreement as to which, if either, treatment is correct. Some reviewers (for example Evans and Langdon, 1976) have accepted Swedlow's analysis whereas others (for example Key, 1969) have questioned the validity of the mathematical analyses. When Poisson's ratio is 0.25, a value close to that for most ceramics, the two criteria are identical and both predict a failure stress independent of σ_2. Griffith's tests on glass tubes and spheres containing single cracks which were fractured under biaxial stress ratios $1 > \sigma_2/\sigma_1 > -1$ supported his own hypothesis, but the total number of specimens was only ten.

Specimens with a random distribution of flaws

In real materials we are not concerned with the stress to fracture a sample with a single flaw but that to propagate whichever flaw, from the statistical distribution available in the material, is subjected to the maximum stress. Our interest lies in determining how the strength observed under uniaxial tension compares with that for other biaxial-stress states. In uniaxial tension only those flaws which are orientated approximately normal to the stress direction are likely to propagate. The flaws parallel to this stress will not propagate and thus the flaws sampled by a unidirectional stress are just a small fraction of the total within the specimen. If, however, the specimen is subjected to an equibiaxial tensile stress then all flaws normal to the plane of the stress are subjected to the same stress. The number of flaws sampled by a biaxial tensile stress is, therefore, much greater than that in the unidirectional case and thus one expects the strength to be lower in the former case. Similarly, the effect of a tensile/compressive stress is also likely to be different from the unidirectional tensile case.

Consider a simple biaxial-stress system with principal stresses σ_1, σ_2, fig. 9.11. The stress in a general direction σ_{12} can be calculated from basic elasticity theory as

$$\sigma_{12} = \sigma_1 \cos^2\phi + \sigma_2 \sin^2\phi \tag{9.33}$$

where ϕ is the angle between the 12-direction and the 1-axis. Weibull theory, as described in § 9.1, has been extended to biaxial stresses for the case of a body of constant volume subjected to a uniform biaxial stress. Essentially the calculation involves substitution of (9.33) into (9.10). Results for Weibull moduli of 5, 10 and 20 are shown in fig. 9.11. The theory predicts that tensile/tensile stresses reduce the fracture strength compared with uniaxial tension, and tensile/compressive stresses the reverse. These effects increase as m decreases and are more pronounced in the tensile/tensile quadrant. With $m = 10$, for example, the equibiaxial strength is $0.8\sigma^*$.

Comparison with experimental data

In view of the inconclusive nature of existing theory, resort to experimental data and empirical approaches is necessary. Measurement of the biaxial strength of ceramics is tedious in that the most convenient test geometry uses a tubular specimen with the stresses applied hydraulically. This test has the particular advantage that the stressed volume can be kept constant whilst the stress ratios are varied, thus eliminating the necessity for stressed-volume corrections. A convenient arrangement is sketched in fig. 9.12. End-caps are cemented on to the specimen ends one of which is free to slide axially in a hydraulic chamber. Three pressures

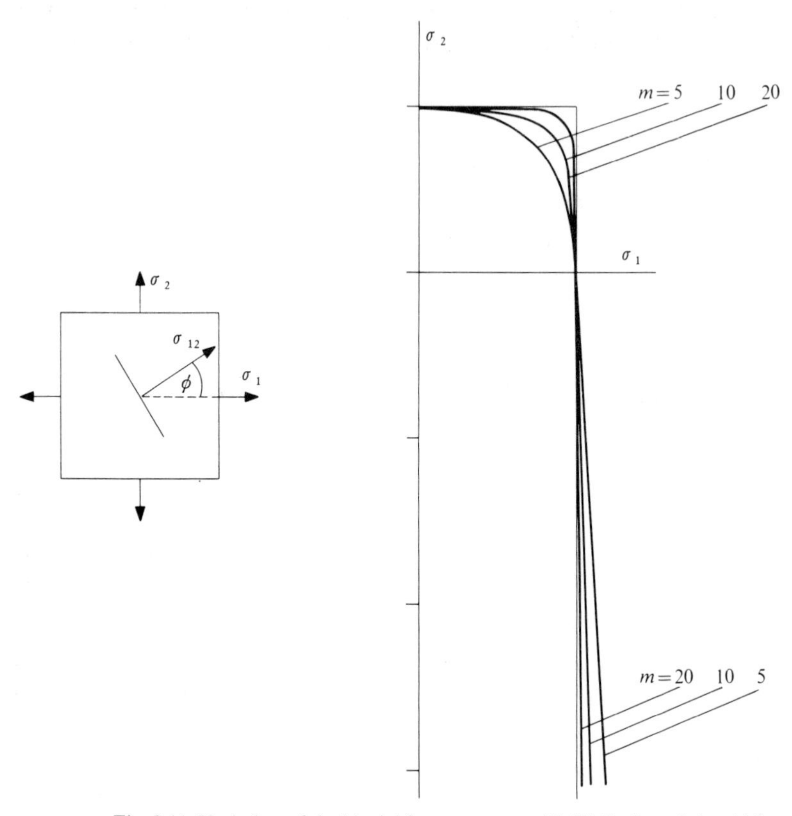

Fig. 9.11. Variation of the biaxial fracture stress with Weibull modulus. (After Salmassy *et al.*, 1955; Price and Cobb, 1972.)

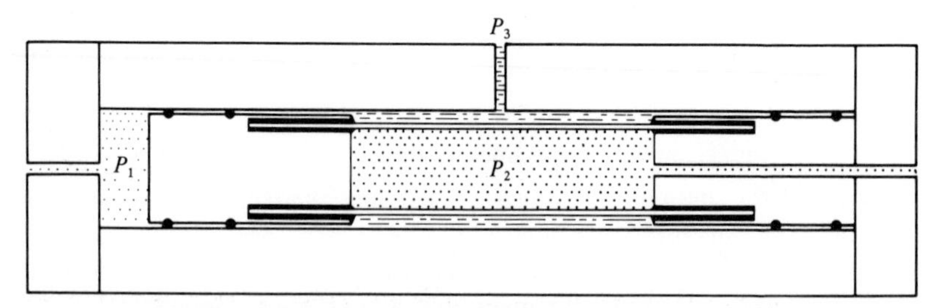

Fig. 9.12. Apparatus for measuring the biaxial strength of ceramic tubes. (After Tappin *et al.*, 1978.)

can be applied independently to the inside, outside and ends of the tube. Application of particular ratios of P_2 with P_1 or P_3 enables a wide range of stress ratios to be obtained.

Available data for several ceramics show that the equibiaxial tensile strength is significantly lower (by 10–20%) than the uniaxial tensile strength. The tensile/compressive strength is generally similar to the

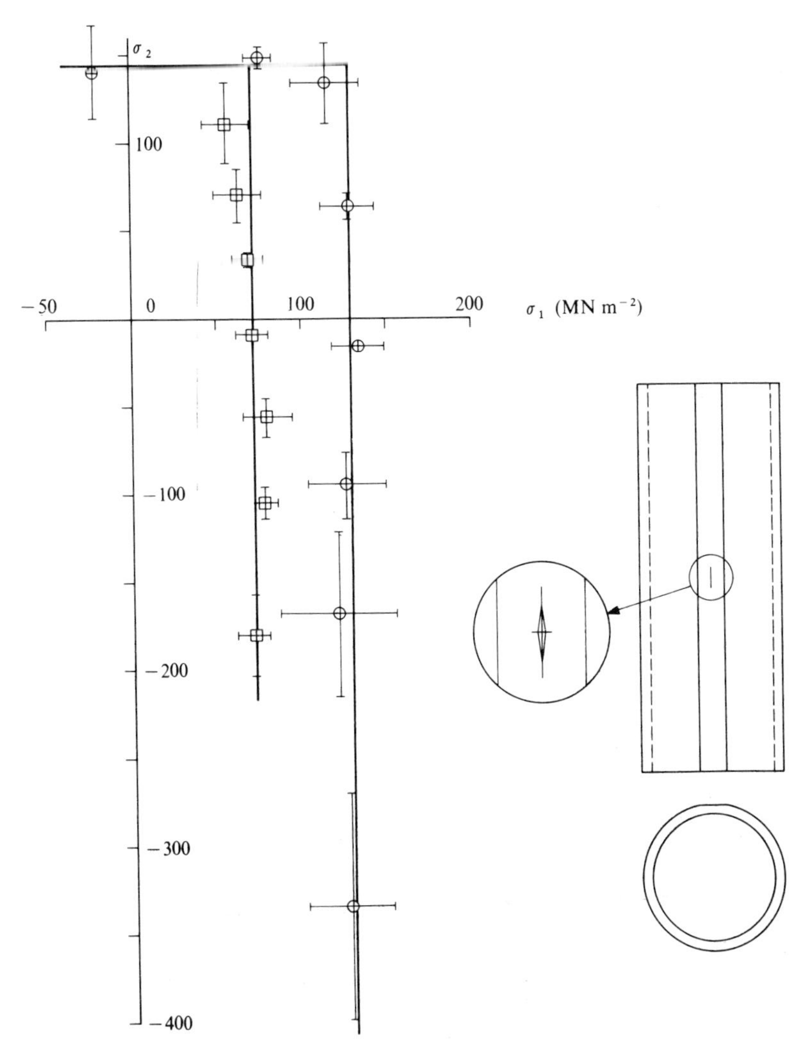

Fig. 9.13. The biaxial strength of self-bonded silicon carbide tubes. o, untreated samples; □, samples with a single large flaw in the orientation shown. (After Tappin *et al.*, 1978.)

uniaxial tensile strength. The data of fig. 9.13 for self-bonded silicon carbide support this general conclusion. Although such effects appear consistent with the Weibull theory shown in fig. 9.11 the second set of data in fig. 9.13 question this severely. Specimens with just a *single* flaw, induced by microindentation, show the same general behaviour as the untreated specimens. The Griffith criterion predicts a strength for specimens with single flaws independent of stress ratio. Poisson's ratio for this SiC is 0.17 and the observed effect is thus opposite to that predicted by Swedlow (1965).

Clearly, further theoretical developments are required. One important experimental observation, for example, that the theories do not recognise, is that the direction of crack propagation on a microscopic scale deviates markedly from the assumed linear path. Most existing theories demand that the original crack is planar and propagates in its own plane. Meanwhile, as a general rule, it would be sensible to assume that the equibiaxial tensile strength of ceramics is conservatively 80% of the uniaxial tensile strength.

References

Wherever possible, acknowledgement has been given in the text to the origin of basic ideas and experimental data, and indications given to sources of background information and to relevant reviews. It is difficult to make more detailed recommendations for further reading in that the requirements of individual readers will vary widely. However, a small number of references are worthy of special mention. The original papers of Griffith (1920), (1924) are still highly readable and have withstood the severest test of all – time. Lawn and Wilshaw (1975) demonstrate how these basic Griffith ideas have been developed to form the current theory of brittle fracture. Much of the necessary background to ceramics, which could only be given brief mention in this volume, is included in the extensive book of Kingery *et al.* (1975). Finally, the review of Evans and Langdon (1976) covers a lot of the present subject matter in a more advanced and detailed manner than space has permitted here.

[*Italic numbers in brackets following each reference indicate the text pages where the reference has been cited.*]

Ashby, M.F. (1972) A first report on deformation mechanism maps. *Acta Met.* **20,** 887. [*74*]

Astbury, N.F. (1963) Mechanical properties of composite materials. In *Advances in Materials Research in the Nato Countries,* p. 369. Advisory Group for Aeronautical Research and Development, Pergamon, London. [*32*]

Aveston, J., Mercer, R.A. and Silwood, J.M. (1975) Fibre reinforced cements – Scientific foundations for specifications. In *Proceedings of the Conference of Composites – Standards, Testing and Design,* p.93. IPC Science and Technology Press, Guildford. [*116*]

Babel, H.W. and Sines, G. (1968) A biaxial fracture criterion for porous brittle materials. *J. Basic Eng.* **90,** 285. [*152*]

Bansal, G.K. (1976) Effect of flaw shape on strength of ceramics. *J. Am. Ceram.Soc.* **59,** 87. [*49*]

Berry, J.P. (1960) Some kinetic considerations of the Griffith criterion for fracture. *J. Mech. Phys. Solids* **8,** 194 and 207. [*46, 47, 129*]

Binns, D.B. (1962) Some physical properties of two-phase crystal-glass solids. *Sci. Ceram.* **1,** 315. [*25, 26*]

Born, M. and Huang, K. (1954) *Dynamical Theory of Crystal Lattices.* Clarendon Press, Oxford. [*22*]

Bowen, D.H., Phillips, D.C., Sambell, R.A.J. and Briggs, A. (1972) Carbon fibre reinforced ceramics. In *Proceedings of the International Conference on Mechanical Behaviour of Materials,* p. 123. Society of Materials Science, Japan. [*112*]

Briggs, A., Bowen, D.H. and Kollek, J. (1974) Mechanical properties and durability of carbon fibre reinforced cement composites. In *Conference on Carbon Fibres, their Place in Modern Technology,* paper 17. Plastics and Rubber Institute, London. [*116*]

Brown, W.F. and Srawley, J.E. (1967) *Plain Strain Crack Toughness Testing of High Strength Metallic Materials,* ASTM Special Technical Publication 410. American Society for Testing Materials, Phila. [*43*]

Bruce, R.H. (1965) Aspects of surface energy of ceramics. *Sci. Ceram.* **2,** 359 and 368. [*76, 77*]

Buerger, J.M. (1930) *Am. Miner.* **15,** 21 and 35. [*52*]

Chalmers, B and Martius, U.M. (1952) Slip planes and the energy of dislocations. *Proc. R. Soc. Lond.* **A213,** 175. [*52*]

Charles, R.J. and Hillig, W.B. (1962) The kinetics of glass failure by stress corrosion. In *Symposium on Mechanical Strength and Ways of Improving it*, p. 511. Union Scientifique Continentale du Verre, Charleroi, Belgium. [*141*]

Chin, G.Y. (1975) Slip and twinning systems in ceramic crystals. In *Deformation of Ceramic Materials* (ed. R. C. Bradt and R. E. Tressler), p. 25. Plenum, New York. [*54, 55*]

Chung, D.H. (1963) Elastic moduli of single crystal and polycrystalline MgO, *Philos. Mag.* **8,** 833.[*22, 26,*]

Clarke, F.J.P. (1964) Residual strain and the fracture stress–grain size relationship in brittle solids. *Acta Met.* **12,** 139. [*84*]

Clarke, F.J.P., Sambell, R.A.J. and Tattersall, H.G. (1962) Mechanisms of microcrack growth in magnesium oxide. *Philos. Mag.* **7,** 393. [*56, 57*]

Coble, R.L. and Kingery, W.D. (1956) Effect of porosity on physical properties of sintered alumina. *J. Am. Ceram. Soc.* **39,** 377. [*27*]

Conrad, H. (1965) Mechanical behaviour of sapphire. *J. Am. Ceram. Soc.* **48,** 195. [*68*]

Cooper, G.A. and Silwood, J.M. (1972) Multiple fracture in a steel reinforced epoxy resin composite. *J. Mater. Sci.* **7,** 325. [*115*]

Davidge, R.W. (1967) The distribution of iron impurity in single crystal MgO and some effects on mechanical properties. *J. Mater. Sci.* **2,** 339. [*69*]

Davidge, R.W. (1969) Mechanical properties of ceramic materials. *Contemp. Phys.* **10,** 105. [*7, 14*]

Davidge, R.W. (1972) The texture of special ceramics with particular reference to mechanical properties. *Proc. Br. Ceram. Soc.* **20,** 364. [*76, 90*]

Davidge, R.W. (1973) Relationships between the microstructure and properties of ceramics. *Sci. Ceram.* **6,** paper IX. [*54, 55*]

Davidge, R.W. (1974) Effects of microstructure on the mechanical properties of ceramics. *Fracture Mechanics of Ceramics* (ed. R. C. Bradt, D. P. H. Hasselman and F. F. Lange), p. 447. Plenum, New York. [*79*]

Davidge, R.W. (1975) The mechanical properties and design data for engineering ceramics. *Ceram. Int.* **1,** 75. [*133*]

Davidge, R.W. (1976) Grain boundaries and the mechanical behaviour of ceramics. In *Grain Boundaries,* p. 41. Institution of Metallurgists, London. [*13*]

Davidge, R.W., Evans, A.G., Gilling, D. and Wilyman, P.R. (1972) Oxidation behaviour of reaction sintered silicon nitride and effects on mechanical properties. In *Special Ceramics,* vol. 5 (ed. P. Popper), p. 107. British Ceramic Research Association, Stoke-on-Trent. [*10, 96, 97*]

Davidge, R.W. and Green, T.J. (1968) The strength of two-phase ceramic–glass materials. *J. Mater. Sci.* **3,** 629. [*86, 87*]

Davidge, R.W., McLaren, J.R. and Tappin, G. (1973) Strength probability time (SPT) relationships in ceramics. *J. Mater. Sci.* **8,** 1699. [*145, 146*]

Davidge, R.W. and Phillips, D.C. (1972) The significance of impact data for brittle non-metallic materials. *J. Mater. Sci.* **7,** 1308. [*108*]

Davidge, R.W. and Tappin, G. (1967) Thermal shock and fracture in ceramics. *Trans. Br. Ceram. Soc.* **66,** 405. [*119, 122, 127*]

Davidge, R.W. and Tappin, G. (1968a) The effective surface energy of brittle materials. *J. Mater. Sci.* **3,** 165. [*41, 49*]

Davidge, R.W. and Tappin, G. (1968b) Internal strain energy and the strength of brittle materials. *J. Mater. Sci.* **3,** 297. [*84, 85*]

Davidge, R.W. and Tappin, G. (1970) The effects of temperature and environment on the strength of two polycrystalline aluminas. *Proc. Br. Ceram. Soc.* **15,** 47. [*45, 81, 98*]

Davidge, R.W., Tappin, G. and McLaren, J.R. (1976) Strength parameters relevant to engineering applications for reaction bonded silicon carbide and REFEL silicon carbide. *Powder Met. Int.* **8,** 110. [*102*]

Davies, D.G.S. (1973) The statistical approach to engineering design in ceramics. *Proc. Br. Ceram. Soc.* **22,** 429. [*139*]

Day, R.B. and Stokes, R.J. (1966) Mechanical behaviour of polycrystalline magnesium oxide at high temperatures. *J. Am. Ceram. Soc.* **49,** 345. [*63*]

Dinsdale, A., Moulson, A.J. and Wilkinson, W.T. (1962) Experiments on the impact testing of cylindrical ceramic rods. *Trans. Br. Ceram. Soc.* **61,** 259. [*109, 110*]

Dinsdale, A. and Wilkinson, W.T. (1966) Properties of whiteware bodies in relation to size of constituent particles. *Trans. Br. Ceram. Soc.* **65,** 391. [*8*]

Eshelby, J.D., Frank, F.C. and Nabarro, F.R.N. (1951) The equilibrium of linear arrays of dislocations. *Philos. Mag.* **42,** 351. [*58*]

Evans, A.G. (1970) Energies for crack propagation in polycrystalline MgO. *Philos Mag.* **22,** 841. [*78*]

Evans, A.G. (1972) Method for evaluating the time-dependent failure characteristics of brittle materials – and its application to polycrystalline alumina. *J. Mater. Sci.* **7,** 1137. [*141*]

Evans, A.G. (1974a) Fracture mechanics determinations. In *Fracture Mechanics of Ceramics* (ed. R. C. Bradt, D. P. H. Hasselman and F. F. Lange), p. 17. Plenum, New York. [*42, 45*]

Evans, A.G. (1974b) High temperature slow crack growth in ceramic materials. In *Ceramics for High Performance Applications* (ed. J. J. Burke, A. E. Gorum *et al.*), p. 373. Brook Hill, Chestnut Hill, Mass. [*142*]

Evans, A.G. (1975) Thermal fracture in ceramic materials. *Proc. Br. Ceram. Soc.* **25,** 217. [*130*]

Evans, A.G. and Davidge, R.W. (1969a) Strength and fracture of fully-dense polycrystalline MgO. *Philos. Mag.* **20,** 373. [*91, 92*]

Evans, A.G. and Davidge, R.W. (1969b) The strength and fracture of stoichiometric polycrystalline UO_2. *J. Nucl. Mater.* **33,** 249. [*93, 94*]

Evans, A.G. and Davidge R.W. (1970) The strength and oxidation of reaction sintered silicon nitride. *J. Mater. Sci.* **5,** 314. [*95, 96, 98*]

Evans, A.G., Gilling, D. and Davidge, R.W. (1970) The temperature dependence of the strength of polycrystalline MgO. *J. Mater. Sci.* **5,** 187. [*72, 92, 93*]

Evans, A. G. and Langdon, T.G. (1976) Structural ceramics. *Prog. Mater. Sci.* **21,** 171. [*54, 55, 65, 143, 152, 157*]

Evans, A.G. and Sharp, J.V. (1971) Microstructural studies on silicon nitride. *J. Mater. Sci.* **6,** 1292. [*95*]

Evans, A.G. and Tappin, G. (1972) Effects of microstructure on the stress to propagate inherent flaws. *Proc. Br. Ceram. Soc.* **20,** 275. [*49, 99*]

Evans, A.G. and Wilshaw, T.R. (1976) Quasi-static solid particle damage in brittle solids – I. Observations, analysis and implications. *Acta Met.* **24,** 939. [*107*]

Evans, R.C. (1964) *An Introduction to Crystal Chemistry*, 2nd edn. Cambridge University Press, London. [*5*]

Fleischer, R.L. (1962) Rapid solid solution hardening, dislocation motion, and the flow stress of crystals. *J. Appl. Phys.* **33,** 3504. [*69*]

Frank, F.C. and Lawn, B.R. (1967) On the theory of Herztian fracture. *Proc. R. Soc. Lond.* **A299,** 291. [*106*]

Garvie, R.C., Hannink, R.H. and Pascoe, R.T. (1975) Ceramic steel? *Nature, Lond.* **258,** 703. [*111*]

Gilman, J.J. (1959) Plastic anisotropy of LiF and other rocksalt-type crystals. *Acta Met.* **7,** 608. [*52, 53*]

Gilman, J.J. (1960) Direct measurements of the surface energies of crystals. *J. Appl. Phys.* **31,** 2208. [*77*]

Gilman, J.J. (1963) The strength of ceramic crystals. In *The Physics and Chemistry of Ceramics* (ed. C. Klinsberg), p. 240. Gordon and Breach, New York. [*24, 28*]

Gilman, J.J. (1974) Theory of solution strengthening of alkali halide crystals. *J. Appl. Phys.* **45,** 508. [*69*]

Glenny, E. and Royston, M.G. (1958) Transient thermal stresses promoted by the rapid heating and cooling of brittle circular cylinders. *Trans. Br. Ceram. Soc.* **57,** 645. [*124, 125, 126*]

Griffith, A.A. (1920) The phenomena of rupture and flow in solids. *Philos. Trans. R. Soc. Lond.* **A221,** 163. [*32, 150, 157*]

Griffith, A.A. (1924) The theory of rupture. In *Proceedings of the First International Congress on Applied Mechanics* (ed. C. B. Biezeno and J. M. Burgers), p. 55. J. Waltman, Delft. [*32, 150, 157*]

Groves, G.W. and Kelly, A. (1963) Independent slip systems in crystals. *Philos. Mag.* **8,** 877. [*61*]

Gumbel, E.J. (1958) *Statistics of Extremes.* Columbia University Press, New York. [*136, 137*]

Hashin, Z. (1968) Elasticity of ceramic systems. In *Ceramic Mircostructures* (ed. R. M. Fulrath and J. A. Pask), p. 313. Wiley, New York. [*24*]

Hasselman, D.P.H. (1969) Unified theory of thermal shock fracture initiation and crack propagation in brittle ceramics. *J. Am. Ceram. Soc.* **52,** 600. [*129, 130*]

Hasselman, D.P.H. and Fulrath, R.M. (1966) Proposed fracture theory of a dispersion-strengthened glass matrix. *J. Am. Ceram. Soc.* **49,** 68. [*89*]

Hasselman, D.P.H. and Fulrath, R.M. (1967) Micromechanical stress concentrations in two-phase brittle-matrix ceramic composites. *J. Am. Ceram. Soc.* **50,** 399. [*88, 89*]

Hertz, H. (1881) *Z. Reine Angew. Math.* **92,** 156. [*105*]

Hill, R. (1952) The elastic behaviour of a crystalline aggregate. *Proc. Phys. Soc. Lond.* **A65,** 349. [*26*]

Hillig, W.B. and Charles, R.J. (1965) Surfaces, stress dependent reactions and strength. In *High Strength Materials* (ed. V. F. Zackay), p. 682. Wiley, New York. [*141*]

Hulse, C.O. and Pask, J.A. (1960) Mechanical properties of magnesia single crystals in compression. *J. Am. Ceram. Soc.* **43,** 373. [*62*]

Inglis, C.E. (1913) Stresses in a plate due to the presence of cracks and sharp corners. *Trans. Inst. Nav. Archit.* **55,** 219. [*35*]

Jaegar, J.C. (1945) Thermal stresses in circular cylinders. *Philos. Mag.* **36,** 418. [*121*]

Johnson, L.G. (1964) *Theory and Technique of Variation Research.* Elsevier, Amsterdam. [*137*]

Johnston, W.G. (1962) Effects of impurities on the flow stress of LiF crystals. *J. Appl. Phys.* **33,** 2050. [*70*]

Johnston, W.G. and Gilman, J.J. (1960) Dislocation multiplication in LiF *Appl. Phys.* **31,** 632.[*56*]

Kelly, A. (1973) *Strong Solids,* 2nd edn. Clarendon Press, Oxford. [*28, 67*]

Kelly, A., Tyson, W.R. and Cottrell, A.H. (1967) Ductile and brittle crystals. *Philos. Mag.* **15,** 567. [*28*]

Key, P.L. (1969) A relation between crack surface displacements and the strain energy release rate. *Int. J. Fract. Mech.* **5,** 287. [*152*]

Kingery, W.D. (1954) Metal–ceramic interactions. *J. Am. Ceram. Soc.* **37,** 42. [*12*]

Kingery, W.D., Bowen, H.K. and Uhlmann, D.R. (1975) *Introduction to Ceramics,* 2nd edn. Wiley, New York. [*2, 5, 157*]

Kronberg, M.L. (1957) Plastic deformation of single crystals of sapphire: basal slip and twinning. *Acta Met.* **5,** 507. [*53, 54*]

Langdon, T.G. (1975) Grain boundary deformation processes. In *Deformation of Ceramic Materials* (ed. R. C. Bradt and R. E. Tressler), p. 101. Plenum, New York. [*73*]

Larson, D.R., Coppola, J.A. and Hasselman, D.P.H. (1974) Fracture toughness and spalling behaviour of high-Al_2O_3 refractories. *J. Am. Ceram. Soc.* **57,** 417. [*128*]

Lawn, B.R. and Wilshaw, T.R. (1975) *Fracture of Brittle Solids.* Cambridge University Press, London. [*37, 106, 143, 157*]

MacKenzie, J.K. (1950) The elastic constants of a solid containing spherical holes. *Proc. Phys. Soc.* **B63,** 2. [*27*]

McLaren, J.R. and Davidge, R.W. (1975) The combined influence of stress, time and temperature on the strength of polycrystalline alumina. *Proc. Br. Ceram. Soc.* **25,** 151. [*101*]

McLaren J.R., Tappin, G. and Davidge, R.W. (1972) The relationship between temperature and environment, texture and strength of self-bonded SiC. *Proc. Br. Ceram. Soc.* **20,** 259. [*11*]

Meredith, H. and Pratt, P.L. (1975) The observed fracture stress and measured values of K_{IC} in commercial polycrystalline aluminas. In *Special Ceramics,* vol. 6 (ed. P. Popper), p. 107. British Ceramic Research Association, Stoke-on-Trent. [*99*]

Meredith, H., Newey, C.W.A. and Pratt, P.L. (1972) The influence of texture on some mechanical properties of debased polycrystalline alumina. *Proc. Br. Ceram. Soc.* **20,** 299. [*100*]

Norton, F.H. (1974) *Elements of Ceramics,* 2nd edn. Addison Wesley, Reading, Mass. [*2*]

Notis, M.R. (1975) Deformation mechanism maps – a review with applications. In *Deformation of Ceramic Materials* (ed. R. C. Bradt and R. E. Tressler), p. 1. Plenum, New York. [*74*]

Nurse, R.W. (1968) Surface energy, adhesion and cohesion in solids. In *Proceedings of the Ninth Conference on the Silicate Industry* (ed. F. Tamas), p. 129. Kultura, Budapest. [*76*]

Nye, J.F. (1957) *Physical Properties of Crystals.* Clarendon Press, Oxford. [*20*]

Orowan, E. (1949) Fracture and strength of solids. *Rep. Prog. Phys.* **12,** 185. [*31*]

Paris, P.C. and Sih, G.C. (1965) Stress analysis of cracks. In *Fracture Toughness Testing and its Applications,* ASTM Special Technical Publication 381, p. 30. American Society for Testing Materials, Phila. [*50*]

Pauling, L. (1948) *The Nature of the Chemical Bond.* Cornell University Press, Ithaca, New York. [*3, 4*]

Petch, N.J. (1968) Metallographic aspects of fracture. In. *Fracture,* vol. 1 (ed. H. Liebowitz), p. 351. Academic Press, New York. [*58*]

Phillips, D.C. (1972) The fracture energy of carbon fibre reinforced glass. *J. Mater. Sci.* **7,** 1175. [*113*]

Phillips, D.C. (1974) Interfacial bonding and the toughness of fibre reinforced ceramics. *J. Mater. Sci.* **9,** 1847. [*116*]

Phillips, D.C., Sambell, R.A.J. and Bowen, D.H. (1972) The mechanical properties of carbon fibre reinforced pyrex glass. *J. Mater. Sci.* **7,** 1454. [*116*]

Pratt, P.L. (1975) Intergranular fracture of polycrystalline ceramics. In *The Mechanics and Physics of Fracture,* paper 22. Institute of Physics, London. [*79*]

Price, R.J. and Cobb, H.R.W. (1972) Application of Weibull statistical theory to the strength of reactor graphite. In *Proceedings of the Conference on Continuum Aspects of Graphite Design* (ed. W. L. Greenstreet and J. C. Battle), p. 547. Oak Ridge National Lab., Tenn. [*154*]

Randall, P.N. (1967) Discussion section of *Plain Strain Crack Toughness Testing of High Strength Metallic Materials* by W. F. Brown and J. E. Srawley, ASTM Special Technical Publication 410, p. 88. American Society for Testing Material, Phila. [*50*]

Reuss, A. (1929) *Z. Angew. Math. Mech.* **9,** 49. [*26*]

Rice, R.W. (1972) Strength and fracture of hot-pressed MgO. *Proc. Br. Ceram. Soc.* **20,** 329. [*80*]

Sack, R.A. (1946) Extension of Griffith's theory of rupture to three dimensions. *Proc. Phys. Soc.* **58,** 729. [*49*]

Salmassy, O.K., Duckworth, W.H. and Schwope, A.D. (1955) *Behaviour of Brittle State Materials,* Technical Report 53-50, part 1. Wright Air Development Centre, Ohio. [*154*]

Sanders, W. (1962) Peierls stress for an idealised crystal model. *Phys. Rev.* **128,** 1540. [*67*]

Simpson, L.A. (1974) Microstructural considerations for the application of fracture mechanics techniques. In *Fracture Mechanics of Ceramics* (ed. R. C. Bradt, D. P. H. Hasselman and F. F. Lange), p. 567. Plenum, New York. [*79, 80*]

Stokes, R.J., Johnson, T.L. and Li, C.H. (1959) The relationship between plastic flow and the fracture mechanism in magnesium oxide single crystals. *Philos. Mag.* **4,** 920. [*60*]

Stokes, R.J., Johnston, T.L. and Li, C.H. (1961) Effect of slip distribution in the fracture behaviour of magnesium oxide single crystals. *Philos. Mag.* **6,** 9. [*57, 58*]

Swedlow, J.L. (1965) On Griffith's theory of fracture. *Int. J. Fract. Mech.* **1,** 210. [*152, 156*]

Tappin, G., Davidge, R.W. and McLaren, J.R. (1978) The strength of ceramics under biaxial stresses. In *Fracture Mechanics of Ceramics* (ed. R. C. Bradt, D. P. H. Hasselman and F. F. Lange), p. 435. Plenum, New York. [*154, 155*]

Vasilos, T. and Passmore, E.M. (1968) Effect of microstructure on deformation of ceramics. In *Ceramic Microstructures* (ed. R. M. Fulrath and J. A. Pask), p. 406. Wiley, New York. [*71*]

Voigt, W. (1928) *Lehrbuch der Kristallphysik,* p. 739. Teubner, Leipzig. [*26*]

von Mises, R. (1928) *Z. Angew. Math. Mech.* **8,** 161. [*61*]

Wachtman, J.B. (1969) Elastic deformation of ceramics and other refractory materials. In *Mechanical and Thermal Properties of Ceramics* (ed. J. B. Wachtman), NBS Special Publication 303, p. 139. National Bureau of Standards, Washington. [*23, 27*]

Weibull, W. (1951) A statistical distribution function of wide applicability. *J. Appl. Mech.* **18,** 293. [*135*]

Wiederhorn, S.M. (1967) Influence of water vapour on crack propagation in soda–lime glass. *J. Am. Ceram. Soc.* **50,** 407. [*142*]

Wiederhorn, S.M. (1969) Fracture of sapphire. *J. Am. Ceram. Soc.* **52,** 485. [*78*]

Wiederhorn, S.M. (1974) Reliability, life prediction, and proof testing of ceramics. In *Ceramics for High Perfomance Applications* (ed. J. J. Burke, A. E. Gorum *et al.*), p. 633. Brook Hill, Chestnut Hill, Mass. [*149*]

Wiederhorn, S.M., Hockey, B.J. and Roberts, D.E. (1973) Effect of temperature on the fracture of sapphire. *Philos. Mag.* **28,** 783. [*78*]

Wiederhorn, S.M., Johnson, H., Diness, A.M. and Heuer, A.H. (1974) Fracture of glass in vacuum. *J. Am. Ceram. Soc.* **57,** 336. [*143*]

Zener, C. (1948) Micromechanism of fracture. In *Fracturing of Metals,* p. 3. American Society of Metals, Cleveland, Ohio. [*58, 59*]

Zwicky, F. (1923) *Phys. Z.* **24,** 131. [*28*]

Index

Numbers in bold face refer to the first page of a chapter or section relating to the subject. The author index is incorporated in the list of references.